DOMESTIC GAS APPLIAN(

CONTENTS

Section 1 — Heating Boilers/Water Heaters

Section 2 — Cookers

Section 3 — Ducted Air Heaters

Section 4 — Fires and Wall Heaters

Section 5 — Tumble Dryers

Section 6 — Instantaneous Water Heaters

Section 7 — Meters

G6

Published by ConstructionSkills, Bircham Newton, King's Lynn, Norfolk, PE31 6RH

© **Construction Industry Training Board 2000**

The Construction Industry Training Board otherwise known as CITB-ConstructionSkills and ConstructionSkills is a registered charity (Charity Number: 264289)

First published 2000
Revised July 2001
Revised October 2001 (Section 4)
Revised December 2002
Revised June 2004
Revised March 2005
Revised October 2006
Revised August 2007

ISBN: 978-1-85751-174-1

ConstructionSkills has made every effort to ensure that the information contained within this publication is accurate. Its content should be used as guidance material and not as a replacement for current regulations or existing standards.

All rights reserved. No part of this publication may be reproduced, stored in a retrieval system or transmitted in any form or by any means, electronic, mechanical, photocopying, recording or otherwise, without the prior permission in writing from ConstructionSkills.

Printed in the UK

For a comprehensive listing of all BES publications turn to the back page.
Tel: 01485 577800 Fax: 01485 577758 E-mail: publications@cskills.org

Heating Boilers/Water Heaters

Data Plate
Family Gas
C E
working on appliances   regulation, 26

CONTENTS

Page   part 9
P48 reg book

	Page
INTRODUCTION	1
The Gas Appliances Directive	1
Gas Appliances (Safety) Regulations 1995	1
European Standards	2
Regulations and Standards Affecting Installation and Maintenance	3
Health and Safety at Work etc. Act 1974 (HSW Act)	3
Management of Health and Safety at Work Regulations 1999 (MHSWR)	3
Reporting of Injuries, Diseases and Dangerous Occurrences Regulations 1995 (RIDDOR 95)	3
Building Regulations 2000 and Building (Scotland) Regulations 2004	3
British Standards	4
Manufacturer's Instructions	5
Specific Gas Controls on Domestic Heating Boilers and Water Heaters	6
Flame Supervision Devices	6
Thermo-electric	6
Testing Thermo-electric Flame Supervision Devices (FSD)	8
Oxygen Depletion System (ODS) – Atmosphere Sensing Device (ASD)	9
Flue Safety Devices	12
Flame Rectification	13
Air Pressure Switch	15
Zero Governor	16
STORAGE WATER HEATERS	18
Integral Storage Heaters	18
Independent Storage Heaters – Directly Heated	20
Independent Storage Heaters – Indirectly Heated	20
Indirect Single Feed (Self-priming) Cylinders	21
Heating Units – Circulators	22
Thermal Store Units	23
Unvented Storage Heaters	23
INSTALLATION OF GAS-FIRED DOMESTIC WATER HEATERS	24
Location Guidelines	25
Large Storage Water Heater Installations	25
Circulator Installations	25
Flues, Ventilation and Fire Precautions	26
Bathrooms and Shower Rooms	27
Living Rooms, Kitchens, Utility Rooms, Halls and Passageways	27
Cloakrooms and Toilets	27
Compartments and Cupboards	27

(continued overleaf)

	Page
Cylinder/Airing Cupboards	28
Under Stairs Cupboards	29
Bedrooms and Bedsitting Rooms	29
Roof Space Installations	30
Back Boiler/Circulator Installations	30
Commissioning Checklist	30
Maintenance Checklist	31
Pre-service Checks	31
Full Service	32
INSTALLATION OF GAS-FIRED CENTRAL HEATING BOILERS	**33**
Types of Appliances	34
System Boilers	34
Heat Exchangers	37
Combination Boilers	38
Condensing Boilers	42
Location Guidelines	44
Flues, Ventilation and Fire Precautions	45
Bathrooms and Shower Rooms	45
Living Rooms, Kitchens, Utility Rooms, Halls and Passageways	45
Compartments and Cupboards	46
Cylinder/Airing Cupboards	46
Under Stairs Cupboards	46
Bedrooms and Bedsitting Rooms	47
Roof Space Installations	47
Commissioning Checklist	48
Open Systems	48
Sealed Systems	49
Maintenance Checklist	49
Pre-service Checks	49
Full Service	50

INTRODUCTION

This introduction contains an appraisal of some of the most important Regulations that determine both the design and manufacture of certain domestic appliances.

To reinforce this, there is reference to the Regulations and British Standards that domestic gas appliance installations must comply with. These Regulations and Standards also impose certain constraints on the subsequent maintenance of appliances and the pipework installation.

THE GAS APPLIANCES DIRECTIVE

Since 1985 the European Union has been pursuing a 'common approach' to tackle the problems associated with trade barriers between member states. For example, EU countries have different laws at present relating to product safety. Removing these barriers is at the heart of the Single European Market introduced in 1992.

The Gas Appliances Directive sets out legal requirements that in future will apply across the European Union. Member countries are required to amend their existing legislation, or to introduce new legislation that conforms with the requirements of the directive. The United Kingdom has implemented the Gas Appliances (Safety) Regulations to conform with the directive.

GAS APPLIANCES (SAFETY) REGULATIONS 1995

Until 1992, the safety (to consumers) of gas appliances sold in the United Kingdom has been covered by the Consumer Protection Act and specifically by the Gas Cooking (Safety) Regulations, and the Heating Appliances (Fire-guards) Regulations. The Gas Appliances (Safety) Regulations 1995 introduced specific requirements.

There was a transitional period until 1996 in which gas appliances offered for sale in the United Kingdom were allowed to meet the old requirements.

The main provision of the new Regulations are:

a) **appliances must be safe**

b) **appliances must be tested**

c) **appliances must be quality guaranteed.**

This means that during the manufacturing process the manufacturer must operate a quality scheme of some type, such as BS 5750, to ensure that all appliances conform to the tested design. This scheme will be monitored by the 'Notified Bodies'.

d) **appliances must carry the CE mark**

All appliances that conform to provisions **a), b) and c)** will carry a CE mark issued by the 'Notified Bodies' (see Figure 1).

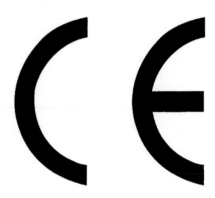

Figure 1 CE mark

The Regulations include detailed procedures for product conformity attestation by third party notified bodies, appointed by the Secretary of State.

All new gas appliances must have information included, covering safe installation, operation and maintenance.

EUROPEAN STANDARDS

European Standards are currently being compiled. For some appliances, where no European Standard is planned, the National Standards (in this country, British Standards) may be recognised. This, for example, will apply to the British type of gas fire.

REGULATIONS AND STANDARDS AFFECTING INSTALLATION AND MAINTENANCE

Health and Safety at Work etc. Act 1974 (HSW Act)

This Act applies to everyone concerned with work activities, ranging from employers, self-employed, and employees, to designers, suppliers and importers of materials for use at work, and people in control of premises. The duties apply both to individual people, and to corporations, companies, partnerships, local authorities etc. Employers have a duty to ensure, so far as is reasonably practicable, the health, safety and welfare at work of all employees, and not to expose people who are not their employees to risks to their health and safety.

Management of Health and Safety at Work Regulations 1999 (MHSWR)

These Regulations impose a duty on employers and self-employed persons to make suitable and sufficient assessment of risks to the health and safety of employees, and non-employees affected by their work. It also requires effective planning and review of protective measures, health surveillance, emergency procedures, information and training.

Reporting of Injuries, Diseases and Dangerous Occurrences Regulations 1995 (RIDDOR 95)

These Regulations require employers to report specified occupational injuries, diseases and dangerous occurrences (events) to the HSE. Certain gas incidents are reportable by suppliers of gas through fixed pipe distribution systems and/or LPG suppliers, and gas installers are required to report certain dangerous gas appliances to the HSE.

Building Regulations 2000 and Building (Scotland) Regulations 2004

These Regulations address the various aspects of building design and construction which include energy conservation and health and safety.

The Secretary of State has approved a number of documents under the Building Regulations 2000 as practical (non-mandatory) guidance to meeting the requirements under the Regulations.

Similar 'deemed to satisfy' guidance is provided in technical handbooks of the Building (Scotland) Regulations 2004.

The documents that particularly relate to gas work in domestic premises are:

- **Building Regulations 2000 (England and Wales)**

    | Part | A | – | Structure |
    | Part | B | – | Fire Safety |
    | Part | F | – | Ventilation |
    | Part | G3 | – | Hot Water Storage |
    | Part | J | – | Combustion Appliances and Fuel Storage Systems |
    | Part | L | – | Conservation of Fuel and Power |
    | Part | M | – | Access To and Use of Buildings |
    | Part | P | – | Electrical Safety |

- **Building (Scotland) Regulations 2004**

    | Section 1 | – | Structure |
    | Section 2 | – | Fire |
    | Section 3 | – | Environment |
    | Section 4 | – | Safety |
    | Section 6 | – | Energy |

BRITISH STANDARDS

British Standards' specifications are an invaluable guide to the installation of gas appliances. If followed, these standards will satisfy the requirements of current Regulations.

The following is a selection of some of the important British Standards Specifications relating to Domestic Gas Appliances, which give guidance on the minimum standard that appliance installations should comply with, to satisfy current Regulations:

BS 5546: 2000	– Specification for installation of gas hot water supplies for domestic purposes
BS 5588: (Domestic) Part 1 1990	– Fire precautions in the design, construction and use of buildings
BS 5864: 2004	– Installation and maintenance of gas fired ducted warm air heaters of rated input not exceeding 70 kW net (2nd and 3rd family gases)
BS 5871: Part 1 2005	– Gas fire, convector heaters and fire/back boilers (2nd and 3rd family gases)

BS 5871:	Part 2 2005	–	Inset fuel effect gas fires of a heat input not exceeding 15 kW (2nd and 3rd family gases)
BS 5871:	Part 3 2005	–	Decorative fuel effect gas appliances of a heat input not exceeding 20 kW (2nd and 3rd family gases)
BS 6700:	1997	–	Design, installation, testing and maintenance of water supplies for domestic purposes
BS 8423:	2002	–	Fire-guards for fires and heating appliances for domestic use - Specification
BS 6172:	2004	–	Installation and maintenance of domestic gas cooking appliances (2nd and 3rd family gases) – Specification
BS 6798:	2000	–	Specification for installation of gas fired boilers of rated input not exceeding 70 kW net
BS 6891:	2005	–	Installation of low-pressure gas pipework of up to 35 mm (R 1¼) in domestic premises (2nd family gases)
BS 5440:	Part 1 2000	–	Flues
BS 5440:	Part 2 2000	–	Air supply
BS 5482:	Part 1 2005	–	Code of practice for domestic butane and propane gas burning installations. Installations in permanent buildings, residential park homes and commercial premises up to 28 mm
BS 7624:	2004	–	Installation and maintenance of domestic direct gas fired tumble dryers up to 6 kW heat input (2nd and 3rd family gases) - Specification

MANUFACTURER'S INSTRUCTIONS

Manufacturer's instructions are important for the installation, commissioning, maintenance and use of any gas appliance. These instructions must be read and followed.

After installation of the appliance or subsequent maintenance of it, the instructions must be returned to the consumer so that they may store them for future reference. This includes both user and installation/servicing instructions. (This is a requirement of the Gas Safety Regulations.)

SPECIFIC GAS CONTROLS ON DOMESTIC HEATING BOILERS AND WATER HEATERS

FLAME SUPERVISION DEVICES

Thermo-electric

In its simplest form, the thermo-electric device (thermocouple) is a loop of two dissimilar metals joined together at one end, with the other ends connected to an electro-magnet. When the joint or junction is heated, a small voltage is produced (see figure 2).

The voltage produced is dependent on the temperature and the metals used. Generally thermocouples used as flame supervision devices utilise a chrome-nickel alloy and copper. The output voltage produced for these metals is between 15 to 30 mv. When the joint is heated by a pilot flame, the voltage energises the magnet thus holding the armature to it in a spring-operated gas valve and allowing gas to flow to the main burner.

Should the pilot be extinguished, the thermocouple would cool down and stop producing a voltage, thus allowing the spring to close the valve.

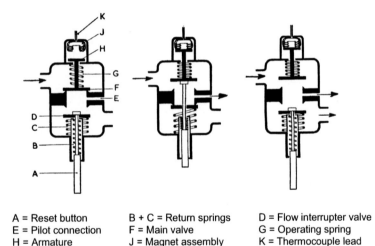

A = Reset button   B + C = Return springs   D = Flow interrupter valve
E = Pilot connection   F = Main valve   G = Operating spring
H = Armature   J = Magnet assembly   K = Thermocouple lead

Figure 2 Thermo-electric flame supervision device

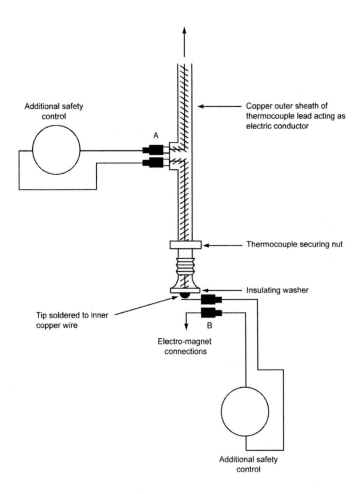

Figure 3 Interruptible thermocouple
(connections made at either position A or B)

Faults

- **Pilot flame**
  Partially blocked or incorrect position of pilot flame, which should play on the tip or the top 12 mm of the thermocouple. Under-aerated pilot flames are easily disturbed by draughts and have a low flame temperature.

- **Thermocouple**
  Contacts must be clean and tight (tight means about a quarter of a turn beyond hand tight). Short circuits can be caused by distortion of insulating washer by overtightening of the union nut. The tip should be clean and undamaged. The tip should be kept below red heat if the thermocouple is to have a reasonable life.

- **Electromagnet**
  Failure rarely occurs of the armature and magnet, which are housed within a sealed unit. Unit exchange is required when this happens.

Testing Thermo-electric Flame Supervision Devices (FSD)

There are numerous procedures for determining whether these devices have shut down correctly. The factors which determine which procedure to adopt are:

- the availability of pressure testing nipples on the equipment involved
- the accessibility of the burner or pilot burner being controlled (i.e. open or room-sealed flues)
- whether the FSD shut-off can be proved by including the procedure within a tightness test of the complete gas installation using the test nipple on the gas meter.

Where the manufacturer's instructions do not specify a procedure then the following may be used:

1. Ensure that all primary and secondary controls are set so that the burner will not be turned off during this procedure.
2. Light the appliance, and allow the burner to reach its normal working temperature.
3. Turn off the appliance shut-off device (appliance isolation valve) and simultaneously start a stop watch.
4. Halt the stop watch when the valve is heard to close.
5. Immediately check that the valve in the FSD has shut off completely, using the most appropriate method as indicated below:

   - **Preferred option**
     Where the appliance gas control system has a test nipple upstream of the FSD device (normally defined as test point P1), test for tightness and let-by between the appliance isolation valve and the FSD. If the FSD device has not shut off completely then a drop in pressure will occur.

- **Option 2**
  Where it is possible to complete a tightness test at the meter, or other suitable position upstream of the appliance isolation valve, then by turning on the appliance isolation valve the integrity of the FSD is also included. (Procedures must be adopted to ensure that any escape indicated by the gauge is not elsewhere on the gas installation pipework other than the FSD.)

- **Option 3**
  Where neither of the above options are available and the appliance is open flued, or the pilot and main burner are readily accessible, then where possible connect a gauge to the burner test nipple and turn on the appliance isolation valve. If any apparent increase in pressure is observed immediately turn off the isolation valve as this indicates that the FSD has not shut off completely. If no apparent increase in pressure is observed (or no gauge has been connected) immediately check with a lighted taper that gas has been interrupted to the main and pilot burner.

6. Check that the time recorded by the stop watch conforms with the current requirements for gas appliances of heat inputs below 60 kW.

Oxygen Depletion System (ODS) – Atmosphere Sensing Device (ASD)

The European Gas Directive, 1 January 1996, states that when undergoing type testing to obtain the CE mark, appliances connected to a flue for the dispersal of combustion products must be so constructed that in abnormal draught conditions there is no release of combustion products in a dangerous quantity into the room concerned.

Domestic gas appliance design allows for excess air under normal operating conditions to be entrained into the appliance combustion chamber and hence to the atmosphere via the flue. When there is a spillage of combustion products into the room where the appliance is installed, complete combustion will occur for a period even though the oxygen level is decreasing and the carbon dioxide level is rising. However, as the oxygen level falls further, incomplete combustion occurs and carbon monoxide (CO) starts to be produced. The appliance design is such that the rate of CO production is initially low as the oxygen level falls and it is at this point that the oxygen depletion system (ODS) within the appliance intervenes.

A typical ODS (S.I.T. Gas Controls Limited) uses a controlled flame to heat a thermocouple, being part of a thermo-electric flame supervision device. As the oxygen level decreases in the atmosphere, so this controlled flame 'lifts' in search of oxygen, thus reducing the heat applied to the tip of the thermocouple until at a pre-determined point the electric current is reduced sufficiently to shut off the gas supply to the appliance (figures 4, 5 and 6).

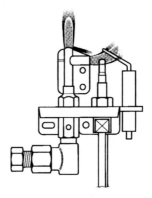

Figure 4 Adequate oxygen supply

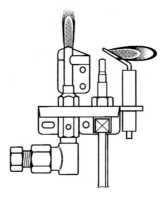

Figure 5 As the oxygen level falls, the sensing flame lifts away from the thermocouple tip

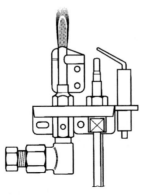

Figure 6 Just prior to shutdown – the sensing flame has completely extinguished

The ODS has an intervention level of 200 ppm (0.02%) of CO concentration in the room in which the appliance is installed.

The installation and annual servicing of all appliances must be conducted by competent operatives and those checks and tests to prevent incomplete combustion occurring as specified in Regulation 26(9) of the current Gas Safety (Installation and Use) Regulations must be complied with.

The ODS device must be checked according to the manufacturer's instructions whenever work has been carried out on the appliances in addition to any annual safety checks. The main points of these checks are to ensure:

- no part of the ODS is damaged
- the ODS is securely mounted in its recommended location
- the flame picture is not distorted and is burning correctly at the main burner cross-ignition port, the sensor port and its inter-connecting ribbon burner
- the aeration port adjacent to the injector is free from any obstruction.

If a customer reports that the ODS keeps 'going out' there is a high probability that it is working correctly and doing exactly what it is supposed to do by making the appliance safe in the event of progressive oxygen starvation due to abnormal flue or ventilation conditions.

Note: The S.I.T. Oxypilot ODS has no serviceable components and if required a complete unit exchange is necessary (figure 7).

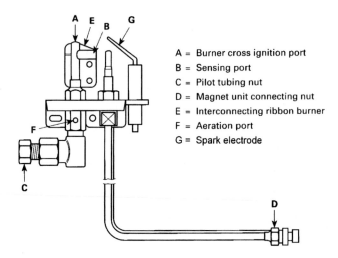

A = Burner cross ignition port
B = Sensing port
C = Pilot tubing nut
D = Magnet unit connecting nut
E = Interconnecting ribbon burner
F = Aeration port
G = Spark electrode

Figure 7 S.I.T. Oxypilot ODS

Flue Safety Devices

These devices are used to detect adverse flue conditions (spillage) at the draught diverter of an open flued appliance. They are known as TTB's (a Dutch acronym of the words 'Themische Tervgslag Beveiliging') but are often referred to as down draught thermostats, thermoswitches or smoke thermostats.

These devices (heat sensors) are located just inside the draught diverter, and are linked:

- in series with the thermocouple of a thermoelectric flame supervision device, or
- to shut off the main burner solenoid valve, or
- to the electronic circuit board.

The sensors are pre-set and calibrated to avoid nuisance shutdowns while still maintaining safe tolerances. They require manual intervention to re-establish the gas supply to the main burner.

Flame Rectification   Ionisation

This method of flame supervision superseded the more basic flame conductance system, which was prone to simulated d.c. flame signals. Condensation or a build up of carbon, due to flame chilling, can bridge the probe and burner. With a d.c. signal where electrons travel around the circuit in only one direction, a control unit can not distinguish the presence of a flame, from the bridging of the gap between the probe and the burner. However an a.c. signal can, due to the two directional flow to and from the control unit.

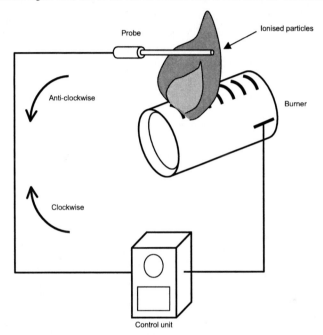

Figure 8 Flame rectification circuit

If we imagine the electrons from the control unit's signal, travelling in a clockwise direction, when the signal reaches the probe, the electrons are able to travel to the burner due to the ionised particles in the gas flame. If there was no flame present, the electrons are not supplied with sufficient pressure to jump the gap e.g. voltage/spark. The probe passing the electrons to the burner is very much like a shotgun firing pellets at a barn door (the burner has a much greater area). Therefore all of the electrons will travel the gap and be registered at the end of the clockwise journey back to the control unit.

When travelling back in the anti-clockwise direction, the electrons now try to pass the gap from the burner to the probe, this is now like shooting a cannon at a pencil, only some of the electrons are able to 'hit' the probe and travel back to the control unit. We now find a rectified signal recognised by the control unit as the presence of a flame.

The flame rectification system can distinguish various signals, for example:

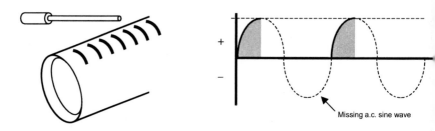

Figure 9 Open circuit

Figure 9 shows the signal read by the control unit where no flame or bridge is present. The electrons reach the end of the probe but have nowhere to go.

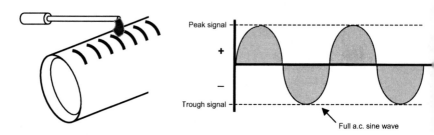

Figure 10 Closed circuit

Figure 10 shows the signal read by the control unit when the gap between the probe and burner, is bridged by conductive matter (condensation or carbon). The electrons can travel freely in both directions.

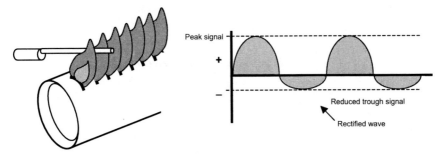

Figure 11 Rectified signal

Figure 11 shows the signal read by the control unit when the gap between the probe and burner, is bridged only by the flame, the shot gun and barn door effect now takes place, rectifying the signal.

A typical, burner head to probe ratio of 8:1 will allow a rectified signal to be produced. This ratio is easily surpassed for most atmospheric bar type burners.

AIR PRESSURE SWITCH

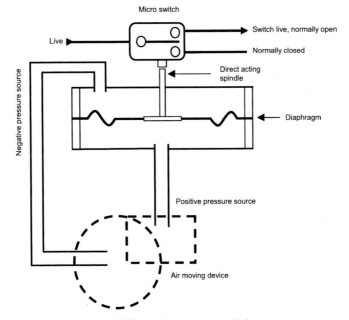

Figure 12 Differential pressure switch

Figure 12 shows a pressure switch, which is sensitive to a pre-set differential pressure. This pressure switch may have a spring adjustment set up, to allow calibration to the desired differential.

ZERO GOVERNOR (equal pressures) equalibrium

On some burners, where gas is to be entrained by a stream of air under pressure, the pressure of the gas must be reduced to atmospheric or 'zero gauge' pressure. Entrainment is achieved by an air blast system's venturi effect.

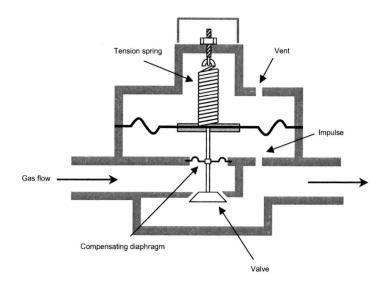

Figure 13 Zero governor

This control is similar to the constant pressure governor, except that instead of a strong compression spring, a spring under tension (pulling force) is fitted to the diaphragm/spindle. The spring supports the weight of the moving parts and is pulled against when the air blast system is in operation, thus opening the valve.

Any increase in pressure above atmospheric, downstream of the valve will cause the valve to close, whereas any drop in outlet pressure below atmospheric will result in the valve being opened.

Zero governors used on air blast systems are generally found downstream of a constant pressure governor.

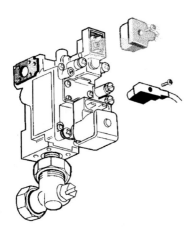 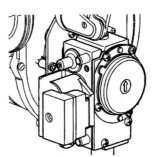

The illustrations above show typical multi-functional valves incorporating zero governor (gas-air proportionating valve) used in conjunction with fan assisted combustion.

In general, the combustion fan will draw air in through a venturi mounted on the fan casing.

When the fan is running it will create suction on the outlet of the gas valve attached to this venturi, such that when the valve is energised gas is drawn out of the valve. In this method, the burner will be inherently safe, as when there is no air there will be no suction and therefore no gas.

Some systems incorporate a feedback tube (impulse line) connected between the air inlet filter/gauze and the top of the zero governor. In the event of the air inlet to the fan becoming restricted, due to linting of the filter/gauze, suction in the tube will increase increasing the suction on the diaphragm to something greater than that of the spring setting and causing the valve to move towards the closed position reducing the gas throughput proportionally.

The amount of gas delivered to the burner from the gas valve will be dependent upon the fan speed and throttle setting on the venturi. If the fan is driven by a simple single speed motor then only adjustment to the gas volume will affect the CO_2 emissions from combustion.

Manufacturer's will usually preset and seal the governor to give the correct air gas ratio. Monitoring the CO_2 emissions in the flue gas will confirm that the system is set correctly for a given gas input pressure.

STORAGE WATER HEATERS

As the name implies, these appliances heat water for storage either in a hot water vessel contained within the appliance, or are connected by flow and return pipes to a separate storage vessel such as a copper cylinder. Both types of unit utilise thermostatic control to restrict the domestic hot water temperature to a maximum of 60°C (140°F).

These appliances are capable of providing a complete hot water service and offer an alternative to instantaneous water heaters. They can be located in positions not available to instantaneous machines (e.g. behind space heaters) and the flow rate delivered to the draw-off taps is generally higher. However, they have the disadvantage of heating water when it is not necessarily required.

There are many designs of storage water heater. The most common are:

- integral storage heaters
- independent storage heaters
- thermal storage heaters
- unvented storage heaters.

INTEGRAL STORAGE HEATERS

These appliances are designed and sized to provide hot water for either domestic or commercial applications and are normally supplied from a cold water cistern.

They have a gas burner which directly heats the water storage vessel (Figure 14) and are normally open flued and protected by a thermoelectric flame supervision device and a thermostat.

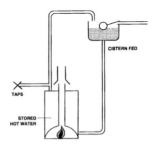

Figure 14 Integral storage unit

The thermostat may be bi-metal snap action or vapour pressure and may be fitted in either the top or the bottom of the hot water storage cylinder.

Figure 15 illustrates the components to be found on a typical appliance.

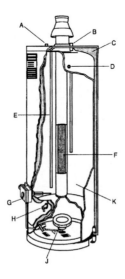

A   Hot water outlet and vent connection to the cistern
B   Cold water inlet from the cistern
C   Insulation
D   Temperature relief valve
E   Sacrificial anode fitted within the storage water to protect the vessel against corrosion

F   Flueway and baffle
G   Multifunctional valve contains thermostat and flame supervision device
H   Drain plug
J   Burner assembly
K   Hot water storage vessel

Figure 15 Cross section through a typical storage heater

In the event that stored water becomes overheated, due to the gas not being shut off by the thermostat, a vent pipe is fitted from the top of the storage unit which discharges into the cistern.

BS 6700 gives recommendations for the design and installation of water supplies and the local water byelaws should also be consulted.

INDEPENDENT STORAGE HEATERS – DIRECTLY HEATED

In this system the water drawn from the storage cylinder has passed directly through the heat exchanger of the appliance.

A typical direct system is shown in Figure 16. Although fairly common, direct systems are not recommended where scale in hard water areas is a problem.

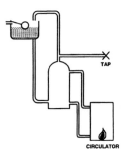

Figure 16 Direct system

INDEPENDENT STORAGE HEATERS – INDIRECTLY HEATED

In an indirect system (Figure 17) the stored water is heated by a calorifier (heat exchanger) within the storage cylinder. This calorifier can be either a coil type or an immersion type as shown in Figure 18.

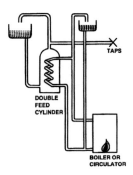

Figure 17 Indirect system

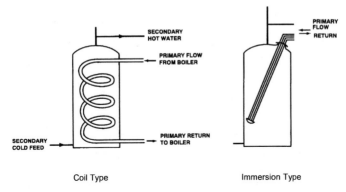

Figure 18 Indirect cylinders and calorifiers

The heat exchanger is connected by flow and return pipes to the heating appliance and there is no direct contact between the water in the appliance and the water drawn from the taps. This system is particularly suitable in hard water areas, as the water circulating through the heat exchanger is not continually replaced with fresh, thus minimising formation of scale.

INDIRECT SINGLE FEED (SELF-PRIMING) CYLINDERS

In the self-priming system (i.e. primatic) both the heating circuit and storage water are fed from one source. There are various single feed methods on the market, one of which is shown in Figure 19. An air lock within the cylinder, formed initially when the system is filled, prevents the heating circuit water from coming into contact with the stored heated water.

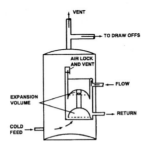

Figure 19 Indirect single feed (self-priming) cylinder

HEATING UNITS – CIRCULATORS

The heating units for independent storage systems are referred to as 'circulators' and may be fitted independently in a kitchen or airing cupboard. They can also be incorporated within a warm air heater unit or located behind a space heater (Figure 20).

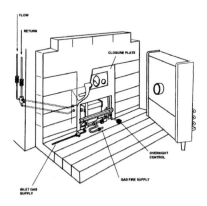

Figure 20 Back circulator installation

The heat exchanger of a circulator is constructed of copper tubes and fins as shown in Figure 21.

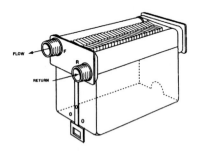

Figure 21 Heat exchanger

THERMAL STORE UNITS

This type of unit works in the reverse order to the indirect cylinder in that the stored hot water heats the water for the draw-off taps as it passes through the storage unit (Figure 22).

Hot water passes from the heat exchanger to the storage unit by gravity or is assisted by a pump. A cistern supplies the water contained within the storage unit. A coil, connected to the mains water supply, is located within the storage unit and delivers domestic hot water via a thermostatic mixing valve. This system may also be utilised to supply an independently pumped central heating circuit.

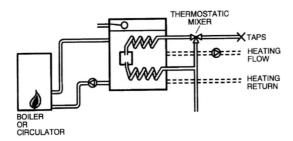

Figure 22 Thermal store unit

UNVENTED STORAGE HEATERS

In this system the storage water is fed directly from the mains, without the need for a cistern or open vent pipe.

To ensure the safety of the system, and prevent contaminated water re-entering the main supply, additional controls are fitted.

The storage water may be directly heated, from a burner which heats the surface of the storage unit, or indirectly heated from a remote heat source. In the indirect system the heat is transmitted to the stored water cylinder via a heat exchanger. The heat source may also supply a central heating circuit.

Local water authorities and the Building Inspectorate must be notified of any intended installation of one of these units. Only those who have received special training are deemed competent to install or maintain this equipment (Building Regulations 1992 – G3).

Further information on unvented storage water training courses can be obtained from ConstructionSkills.

INSTALLATION OF GAS-FIRED DOMESTIC WATER HEATERS

Advice on materials, design and installation of water heaters is given in BS 5546: Specification for installation of gas hot water supplies for domestic purposes.

Further advice can be found in BS 6700: Specification for design, installation, testing and maintenance of services supplying water for domestic use within buildings and their curtilages.

Any work that is carried out on water installations in the United Kingdom must conform to the Water Supply Byelaws. These byelaws cover areas relevant to the installation of hot water appliances such as:

- acceptable materials used for drinking water supplies
- protection against contaminated water entering the main water supply
- cold water storage protection
- identification of water pipes
- unvented hot water systems.

Before installing any gas appliance, the manufacturer's instructions should be consulted. They give specific directions for the installation of the appliance, including location requirements and minimum dimensions with regard to flue terminal positions.

For some appliances you may find the measurements given differ slightly from those stated in the British Standards. In this case the manufacturer has designed the appliance to operate satisfactorily with regard to specific installation instructions and they should, therefore, be adhered to.

The installation must also conform to the current Gas Safety Regulations and, if the installation contains an electrical supply, to the Institute of Electrical Engineers Regulations (IEE).

The choice of a particular type of system, storage or instantaneous, single point or multipoint, should match the requirements of the customer; taking into account acceptable positions for installation, availability of water and comparative installation and maintenance costs.

LOCATION GUIDELINES

Although the manufacturer's instructions give precise information regarding acceptable locations for appliances, the following points should be considered before selecting the exact location:

- suitability of the site
- route of any flue system
- acceptable terminal position
- available ventilation, if required
- location of appliance to other parts of the system, i.e. storage cylinders and cisterns
- pipework route for water supplies
- pipework route for gas supplies
- suitable electrical supply, if applicable
- access for maintenance and servicing
- protection against frost damage, if required.

LARGE STORAGE WATER HEATER INSTALLATIONS

Where the cold water is supplied from a feed cistern, the height of the cistern should be such that an acceptable flow of water is available at the highest draw-off point. The weight of the appliance, when full of water, is considerable and this should be taken into account when deciding its location and means of support.

CIRCULATOR INSTALLATIONS

The water circulation rate is an important factor in achieving a satisfactory service from a circulator and storage vessel. It is dependent upon the effective circulation head (Figure 23) and the size of the flow and return pipes.

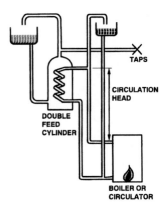

Figure 23 Circulating head

An inadequate circulation head may give rise to overheating and noisy operation of the appliance. The manufacturer will give specific guidance in the installation instructions regarding the minimum circulation head.

FLUES, VENTILATION AND FIRE PRECAUTIONS

An adequate supply of air is essential for a gas appliance to burn safely and efficiently. For flueless and open flued appliances, an air supply is also required to ensure adequate ventilation in the room or space in which the appliance is located.

The air requirement for a given appliance is dependent upon its rated heat input, whether it is flueless, open flued or room sealed, and its location in a room or a compartment. Room sealed water heaters are permissible in any room or internal space provided that the flue terminal can be located as recommended by the manufacturer or BS 5440-1.

In new installations the following water heaters must be flued:

- large storage heaters

- circulators and back boilers

- any water heater supplying a bath.

The following gives guidance on acceptable locations for a water heater with regard to flueing and ventilation.

More detailed information on flueing and ventilation of gas appliances can be found in the ConstructionSkills publication *Gas Safety*.

Bathrooms and Shower Rooms

Under no circumstances shall any open flued or flueless water heater be installed in a room containing a bath or shower.

Only room sealed appliances may be fitted in these locations, provided that the flue terminal can be located as recommended in BS 5440.

Where a room sealed water heater, with an electrical supply, is located in a room containing a fitted bath or shower, the mains electrical switch should be so sited that it cannot be touched by a person using the bath or shower, in accordance with the Institute of Electrical Engineers Regulations.

Living Rooms, Kitchens, Utility Rooms, Halls and Passageways

When selecting a living room as a location for a water heater careful consideration should be given to the effect on the living area of such an installation, e.g. noise of the appliance, general aesthetics of the installation and servicing requirements.

The type of appliance that can be fitted is unrestricted providing there is sufficient ventilation and an adequate flue system is available.

Cloakrooms and Toilets

All types of water heater are permissible provided that flueing and ventilation meet the requirements of BS 5440. Flueless water heaters should not be installed where the volume of the room is 5 m^3 or less. Any air vent must communicate directly with the outside air.

Compartments and Cupboards

A compartment is defined as an enclosed space within a building, either constructed or modified specifically to accommodate the water heater and its ancillary equipment.

The regulations do not permit flueless water heaters being installed in compartments.

Where a compartment is used to accommodate a flued water heater the compartment should conform to the following requirements:

- be a rigid structure in accordance with any manufacturer's instructions regarding the internal surfaces

- combustible internal surfaces should be at least 75 mm (3 in) from any part of the heater or be suitably protected

- must have access to allow inspection, servicing and removal of the appliance and ancillary equipment

- the compartment must incorporate air vents for ventilation and, where necessary, for combustion as recommended in BS 5440-2.

- no air vent must communicate with a private garage, bathroom or shower room, bedroom or bedsitting room, if an open flued appliance is fitted in the compartment.

Cylinder/Airing Cupboards

A circulator, with an input rating not exceeding 7 kW (30,600 Btu/h) may be fitted within this area, providing certain requirements are met in addition to those specified for compartments (Figure 24).

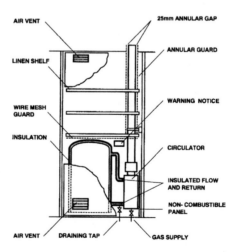

Figure 24 Typical circulator installation in an airing cupboard

A warning label should advise the customer of the potential dangers of storing, drying and airing clothes in the circulator compartment.

Any airing space should be separated from the circulator compartment by a non-combustible material which may have perforation holes, provided they are not greater than 13 mm (½ in) across any axis. This prevents any clothing falling onto the heater.

The flue pipe should not pass through the airing space unless protected by a guard, such as wire mesh, that prevents stored linen being placed within 25 mm (1 in) of the flue pipe.

Any insulating material used on a cylinder fitted adjacent to an open flued circulator shall be made of, or protected by, a non-combustible material.

Under Stairs Cupboards

This location should only be considered as a last resort in a building that is more than two storeys high. Wherever possible only room sealed appliances should be installed. Air vents must be direct to the outside air.

The cupboard should be treated as a compartment and, in addition, in a building that is more than two storeys high all the surfaces should be lined with non-combustible material unless inherently fire resistant, e.g. plastered ceiling.

Bedrooms and Bedsitting Rooms

Water heaters of input more than 14 kW shall be room sealed. Non-room sealed appliances of 14 kW or less, must incorporate a safety control designed to shut-off the appliance before there is a build-up of a dangerous quantity of combustion products in the room concerned.

In addition to room sealed appliances, the following water heaters may be fitted provided the room volume exceeds 20 m^3 and the necessary ventilation requirements can be met:

- flueless storage heater with a storage capacity not exceeding 9 litres (2 gallons) and a heat input not exceeding 3 kW (10,000 Btu/h) serving a wash basin or sink only

- open flued circulator with a heat input not exceeding 4.5 kW (15,300 Btu/h).

When installing a water heater in a bedroom, consideration should be given to the factors affecting amenity, e.g. noise of operation.

Roof Space Installations

Where no alternative exists and where local regulations permit, an open flued or room sealed appliance may be fitted, provided the appliance is protected from frost damage.

When this location is selected the roof space should have:

- flooring of sufficient strength and area to support the appliance and facilitate servicing

- enough vertical clearance, when siting the cold water cistern, to ensure the availability of the static head required by the appliance

- a suitable means of access to the heater, e.g. foldaway loft ladder, and sufficient fixed lighting

- a guard fitted to prevent items stored within the roof space coming into contact with the appliance

- a guard-rail around the access hatch

- accessible gas isolation control outside the loft area.

Back Boiler/Circulator Installations

Where the back boiler or circulator is providing only domestic hot water there is no requirement for a flue liner or terminal to be fitted to the chimney.

Where the back boiler or circulator is used to supply hot water to radiators, a flue liner and terminal must be fitted.

Pipework located within the builder's opening must be protected from corrosion.

When calculating ventilation requirements, the installation is not treated as a combined appliance if it is only supplying domestic hot water.

COMMISSIONING CHECKLIST

It should be noted that the checklist shown may be used in the absence of the appliance manufacturer's instructions.

1. Flush out primary water system at least twice. Flush once cold and once hot.
2. Check all joints for water leaks.
3. Fill the system with water and remove air.
4. Test the installation for gas tightness and purge.

5. Check ventilation is adequate, where appropriate.
6. Check ignition device and light pilot.
7. Check pilot flame length and position and operation of flame supervision device.
8. Check the burner pressure. Re-check after the appliance has been allowed to heat up for about 15 minutes.
9. Check the gas rate, if necessary, and the flame picture.
10. Check for spillage at the draught diverter of open flued appliances.
11. Check all controls are operating satisfactorily.
12. Leave the installation and servicing instructions and the operating instructions with the customer.
13. Instruct the customer on the operation of the appliance and all controls.
14. Instruct the customer on the need for regular servicing of the installation.

MAINTENANCE CHECKLIST

It should be noted that the checklist shown may be used in the absence of the appliance manufacturer's instructions.

Pre-service Checks

1. Check with the customer to ascertain any problems or faults with the installation.
2. Check the general condition of the appliance and that the installation conforms to appropriate standards.
3. Check that there is adequate ventilation and that the flue is both routed and terminated correctly.
4. Check there are no signs of spillage on the appliance or adjacent walls.
5. Check the operation of all controls, the flame supervision device and ignition system.
6. Check flame picture.
7. Check for any signs of water leakage.
8. Advise the customer of any problems.

Full Service

1. Isolate gas, water and where applicable electricity supplies.
2. Dismantle as necessary and clean dust and deposits from within the casing and/or the fireplace opening.
3. Check for signs of damage to electrical connections, cables or components, clean and rectify as necessary.
4. Remove and clean main burner and injectors.
5. Remove and clean pilot burner and injectors.
6. Inspect the heat exchanger, flueways and carry out flue flow test.
7. Ease and grease gas taps as necessary.
8. Check all appliance seals and joints.
9. Check all disturbed gas connections.
10. Test the ignition device by lighting the appliance.
11. Check pilot flame and test flame supervision device.
12. Ensure that the working pressure, gas rate and flame picture are correct.
13. Check operation of controls as fitted, including:
    - appliance thermostat
    - cylinder thermostat
    - frost thermostat
    - overheat thermostat
    - safety valve.
14. Check pipework and valves for water leaks.
15. Check ball-valve in feed and expansion cistern.
16. Check operation of all electrical controls.
17. Check for spillage.
18. Leave the appliance in a working order and advise the customer of any further work required.

INSTALLATION OF GAS-FIRED CENTRAL HEATING BOILERS

Advice on materials, design and installation of hot water boilers is given in BS 6798: Specification for installation of gas-fired hot water boilers of rated net input not exceeding 70 kW.

Further advice can be found in BS 6700: Specification for design, installation, testing and maintenance of services supplying water for domestic use within buildings and their curtilages.

Any work that is carried out on water installations in the United Kingdom must conform to the Water Supply Byelaws. These byelaws cover areas relevant to the installation of hot water appliances such as:

- acceptable materials used for drinking water supplies
- protection against contaminated water entering the main water supply
- cold water storage protection
- identification of water pipes
- unvented hot water systems.

Before installing any gas appliance, the manufacturer's instructions should be consulted. They give specific directions for the installation, including location requirements and minimum dimensions with regard to flue terminal positions.

For some appliances you may find the measurements given differ slightly from those stated in the British Standards. In this case the manufacturer has designed the appliance to operate satisfactorily with regard to specific installation instructions and these should be adhered to.

The installation must also conform to the current Gas Safety Regulations and, if the installation contains an electrical supply, to the Institute of Electrical Engineers Regulations (IEE).

The choice of a particular type of gas-fired hot water boiler, should match the requirements of the customer, taking into account acceptable positions for installation, availability of water and comparative installation and maintenance costs.

TYPES OF APPLIANCES

System Boilers

When the total heat requirements have been calculated, the installer will need to select the boiler. The heat loss calculation will determine the boiler size required.

The installer needs to consider a number of points before selecting a boiler:

- location

- the boiler heat output

- the preference of the customer, for example wall hung, free standing, combination boiler, back boiler or specific manufacturer, etc.

- the physical size of any particular boiler

- whether the customer is likely to extend the property in the future, therefore adding to the total heating requirements

- type of system being installed, or in the case of an existing system, its compatibility

- whether open flued or room sealed.

Having considered all of the above the selection process can be finalised. In essence the boiler should be matched to the total calculated heating and hot water requirements. It can be prudent to select a boiler slightly larger than is required to allow the system to be extended at a later date.

Gas central heating boilers offer a choice of models. They may be either:

- open flued

- room sealed.

Both categories are available in natural draught or fanned draught versions. Boilers can generally be categorised in the following groups:

- system boilers, cast iron heat exchanger   – floor standing   (Figure 25)
       – wall hung   (Figure 26)
       – back boiler   (Figure 27)

- low water content, copper heat exchanger (wall hung)

- combination boilers (either direct or indirect heat exchanger)

- system and combination boilers, high efficiency (condensing).

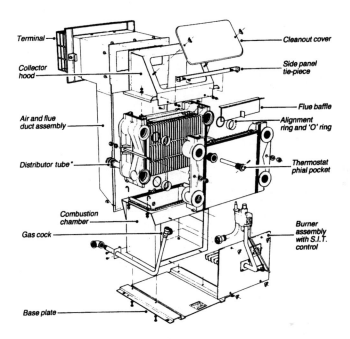

Figure 25 A typical floor standing room sealed boiler with cast iron heat exchanger

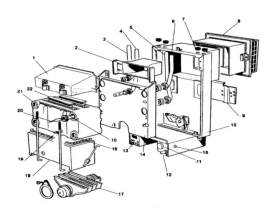

1.	Collector hood	9.	Wall mounting plate	16.	Overheat thermostat (if fitted)
2.	Boiler flue duct	10.	Back panel	17.	Main burner
3.	Gravity flow pipe	11.	Control thermostat	18.	Combustion chamber
4.	Pumped return pipe	12.	Ignitor button	19.	Boiler drain point
5.	Gravity return pipe	13.	Programmer (optional)	20.	Boiler thermostat
6.	Pumped flow pipe	14.	Interpanel	21.	Heat exchanger
7.	Rubber sealing grommets	15.	Thermostat pocket	22.	Flue baffles
8.	Balanced flue terminal				

Figure 26 A typical wall hung room sealed boiler with cast iron heat exchanger

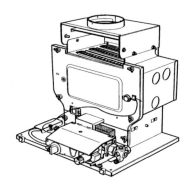

Figure 27 A typical open flued back boiler with cast iron heat exchanger

Heat Exchangers

Cast Iron Heat Exchanger

This type of heat exchanger is available in wall hung, floor standing and back boiler types.

Some of the advantages are:

- quiet operation
- proven track record
- extensive range of models
- compatibility with most types of open vented system design.

Some of the disadvantages are:

- some models can be prone to leakage from gaskets and manifolds
- can be heavy to carry and install. Can be prone to sludge deposits if system not correctly installed.

Low Water Content (Copper Heat Exchanger)

Most widely used in wall hung boilers. The low water content assists in the heat recovery rate of the system, and they have become very popular in recent times.

Some of the advantages are:

- fast transfer of heat through the exchanger
- lighter weight construction
- may be of smaller dimensions
- many are suitable for sealed systems.

Some of the disadvantages are:

- only suitable for pumped primary systems
- heat exchangers can become partially blocked with system debris
- noise
- may require a system bypass.

Combination Boilers

Combination boilers combine the functions of a central heating boiler and an instantaneous multi-point water heater, giving priority to the supply of domestic hot water. The combination boiler has all its operating components contained within the casing (Figure 28a and Figure 28b) and is typically designed for use within a sealed central heating system.

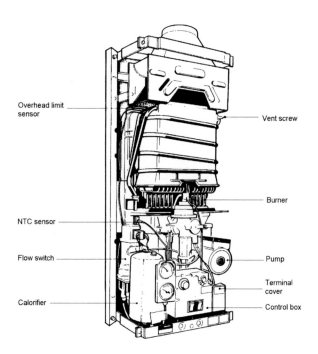

Figure 28a Typical combination boiler

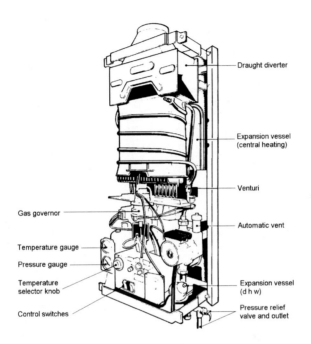

Figure 28b Typical combination boiler

The installation of a combination system is considerably easier than that of a conventional system as the boiler incorporates an expansion vessel. This eliminates the need for a feed and expansion cistern in the roof space. A comparison of the two systems can be seen in Figure 29 below.

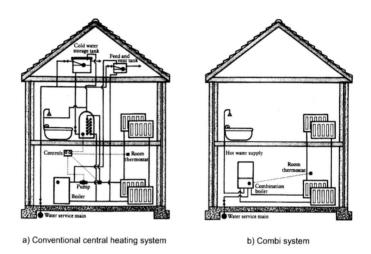

a) Conventional central heating system   b) Combi system

Figure 29 Comparison of heating systems

The combination boiler, just like a conventional system boiler, has a gas-fired burner and a heat exchanger, the waterways of which form part of the central heating system. Hot water from the heat exchanger passes to a diverter valve, which directs the flow to a domestic hot water calorifier or central heating circuit. Figure 30 shows the position of these principal components.

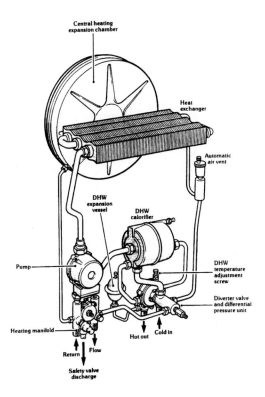

Figure 30 Position of the principal components within a combination boiler

The calorifier is tightly packed with small copper tubes, through which secondary water flows the moment a hot water draw-off is opened. As the water flows it picks up heat from the surrounding primary water in the calorifier. The central heating system is pressurised and fitted with a relief valve to protect against any excess build-up of pressure in the system. A gauge on the appliance indicates the operating pressure. A bypass, connected across the primary flow and return, ensures that an adequate flow of water is maintained. This bypass is sometimes a built-in feature of the appliance.

Condensing Boilers

The condensing boiler can operate at an efficiency of up to 94% by recovering the heat that, in a traditional boiler, escapes up the flue. The temperature of the products of combustion in a traditional boiler ranges from 200–500°C, which prevents condensing of the products in the flue. Up to 25% of the energy contained in the gas escapes as sensible heat and latent heat.

Note: Sensible heat is the heat gained or lost by a substance when its temperature changes, but its state remains the same. Latent heat is the heat gained or given up by a substance when it changes from one physical state to another e.g. from a gas to a liquid.

Adding extra heat exchanger area lowers the flue temperature, encouraging the water vapour to condense, and recovers both sensible and latent heat which would usually be lost in the flue.

Condensing boilers have either a stainless steel or aluminium alloy heat exchanger, as these are able to resist attack from the slightly acidic condensate.

The condensate which can run to as much as 2 litres a day, is drained off from the bottom of the heat exchanger.

Boiler layout varies from manufacturer to manufacturer. Figure 31 shows two typical designs.

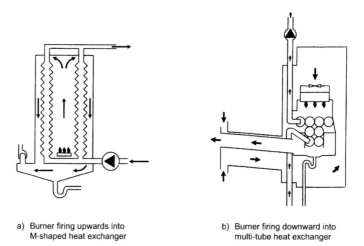

a) Burner firing upwards into M-shaped heat exchanger

b) Burner firing downward into multi-tube heat exchanger

Figure 31 Two condensing boiler options

Figure 31a shows a burner firing upwards into a cast heat exchanger which forms an up and over configuration. Heat transfer is such that condensation is only formed on the downward pass.

The design in Figure 31b consists of a multi-tube arrangement. The fan assisted burner fires downward through the heat exchanger. The condensate forms on the lower tubes and runs to a drain at the bottom of the appliance.

Condensate Discharge Arrangements

Condensate produced within the appliance and flue should preferably be drained into the house sanitary waste system, or alternatively the rainwater system. The plastic discharge pipework, minimum diameter 22 mm, can terminate at a branch or stack internal to the house, or externally at an open gully and must be fitted with a trap comprising of a 75 mm condensate seal. If neither of these arrangements is practicable the condensate can be discharged into a purpose-made soakaway.

Externally run condensate pipe can be prone to freezing. To avoid this, as much pipework as possible should be internal and, after passing through the wall, terminated into an external drain using the minimum number of joints and bends (see Figure 32).

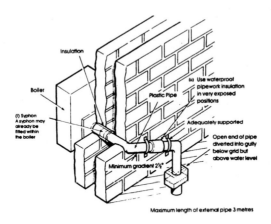

Figure 32 Condensate pipework arrangement to prevent freezing

Installation

A condensing boiler requires a well-designed central heating and hot water system to fully exploit its efficiency and permit a temperature differential between flow and return of about 20°C. Radiators installed to a condensing boiler system need to be oversized by approximately 10% due to their lower average temperature.

When replacing a conventional central heating boiler with a condensing boiler it is important to confirm that the existing radiator outputs are satisfactory. It may also be necessary to install an additional radiator in the room that housed the existing boiler if the boiler provided a source of heat for the room.

LOCATION GUIDELINES

Although the manufacturer's instructions give precise information regarding acceptable locations for appliances, the following points should be considered before selecting the exact location:

- suitability of the site
- route of any flue system
- acceptable terminal position
- available ventilation, if required
- location of appliance to other parts of the system, i.e. storage cylinders and cisterns
- pipework route for water supplies
- pipework route for gas supplies
- suitable electrical supply, if applicable
- access for maintenance and servicing
- protection against frost damage, if required.

FLUES, VENTILATION AND FIRE PRECAUTIONS

An adequate supply of air is essential for a gas appliance to burn safely and efficiently. For open flued appliances, an air supply is also required to ensure adequate ventilation in the room or space in which the appliance is located.

The air requirement for a given appliance is dependent upon its rated heat input, whether it is open flued or room sealed, and its location in a room or a compartment. Room sealed hot water boilers are permissible in any room or internal space provided that the flue terminal can be located as recommended by the manufacturer or BS 5440-1.

All fan-assisted flue systems must incorporate a fan pressure switch which automatically turns off the appliance if the draught falls below a preset minimum standard.

The following gives guidance on acceptable locations for a hot water boiler with regard to flueing and ventilation.

More detailed information on flueing and ventilation of gas appliances can be found in the ConstructionSkills publication *Gas Safety*.

Bathrooms and Shower Rooms

Under no circumstances shall any open flued hot water boiler be installed in a room containing a bath or shower.

Only room sealed appliances may be fitted in these locations, provided that the flue terminal can be located as recommended in BS 5440.

Where a room sealed hot water boiler, with an electrical supply, is located in a room containing a fitted bath or shower, the mains electrical switch should be so sited that it cannot be touched by a person using the bath or shower, in accordance with the Institute of Electrical Engineers Regulations.

Living Rooms, Kitchens, Utility Rooms, Halls and Passageways

When selecting a living room as a location for a hot water boiler careful consideration should be given to the effect on the living area of such an installation, e.g. noise of the appliance, general aesthetics of the installation and servicing requirements.

The type of appliance that can be fitted is unrestricted providing there is sufficient ventilation and an adequate flue system is available.

Compartments and Cupboards

A compartment is defined as an enclosed space within a building, either constructed or modified specifically to accommodate the hot water boiler and its ancillary equipment.

Where a compartment is used to accommodate a hot water boiler the compartment should conform to the following requirements:

- be a rigid structure in accordance with any manufacturer's instructions regarding the internal surfaces

- combustible internal surfaces should be at least 75 mm (3 in) from any part of the heater or be suitably protected with a non-combustible material

- must have access to allow inspection, servicing and removal of the appliance and ancillary equipment

- the compartment must incorporate air vents for ventilation and, where necessary, for combustion as recommended in BS 5440-2.

- no air vent must communicate with a private garage, bathroom or shower room, bedroom or bedsitting room, if an open flued appliance is fitted in the compartment.

Cylinder/Airing Cupboards

A warning label should advise the customer of the potential dangers of storing, drying and airing clothes in the boiler compartment.

Any airing space should be separated from the boiler compartment by a non-combustible material which may have perforation holes, provided they are not greater than 13 mm (½ in) across any axis. This prevents any clothing falling onto the heater.

The flue pipe should not pass through the airing space unless protected by a guard, such as wire mesh, that prevents stored linen being placed within 25 mm (1 in) of the flue pipe.

Under Stairs Cupboards

This location should only be considered as a last resort in a building that is more than two storeys high. Wherever possible only room sealed appliances should be installed. Air vents must be direct to the outside air.

The cupboard should be treated as a compartment and, in addition, in a building that is more than two storeys high all the surfaces should be lined with non-combustible material unless inherently fire resistant, e.g. plastered ceiling.

Bedrooms and Bedsitting Rooms

Gas fired hot water boilers should only be installed in a room used or intended to be used as sleeping accommodation if:

- it is a room sealed appliance

- when not room sealed, it is 14 kW or less and incorporates a device that will shut off the gas supply before dangerous levels of products of combustion build up.

Roof Space Installations

Where no alternative exists and where local regulations permit, an open flued or room sealed appliance may be fitted, provided the appliance is protected from frost damage.

When this location is selected the roof space should have:

- flooring of sufficient strength and area to support the appliance and facilitate servicing

- enough vertical clearance, when siting the cold water cistern, to ensure the availability of the static head required by the appliance

- a suitable means of access to the heater, e.g. foldaway loft ladder, and sufficient fixed lighting

- a guard fitted to prevent items stored within the roof space coming into contact with the appliance

- a guard-rail around the access hatch

- accessible gas isolation control outside the loft area.

COMMISSIONING CHECKLIST

It should be noted that the checklist shown may be used in the absence of the appliance manufacturer's instructions.

Open Systems

1. With all valves in an open position, flush out water system at least twice. Flush once cold and once hot.
2. Check all joints for water leaks.
3. Fill the system with water and remove air.
4. Add the corrosion inhibitor.
5. Test the installation for gas tightness and purge.
6. Check ventilation is adequate, where appropriate.
7. Check ignition device and light pilot.
8. Check pilot flame length and position and operation of flame supervision device.
9. Check the burner pressure. Re-check after the boiler has been allowed to heat up for about 15 minutes.
10. Check the gas rate, if necessary, and the flame picture.
11. Adjust the pump to give the required design temperature difference between flow and return pipes.
12. Check for noisy system operation.
13. Check for pumping over at open vent.
14. Balance the system to give the correct design temperature drop.
15. Check for spillage at the draught diverter of open flued boilers.
16. Check all controls are operating satisfactorily.
17. Leave the installation and servicing instructions and the operating instructions with the customer.
18. Instruct the customer on the operation of the boiler system and all controls.
19. Instruct the customer on the need for regular servicing of the installation.

Sealed Systems

Commission in a similar way to open systems and include the following:

1. Fill the system from the mains water supply with the use of an approved filling loop or by a sealed system filler pump and break tank.

2. Following the flushing and refilling of the system:

    - introduce or release water until the desired pre-pressurisation level is achieved, or
    - drain water from the safety valve, allowing the level in the top up bottle to fall, then refill the bottle.

3. Ensure that the safety valve operates at a pressure of 3 bar, by noting the reading on the pressure gauge (allow ± 0.3 bar).

4. Ensure that the boiler high temperature cut-off operates correctly.

MAINTENANCE CHECKLIST

It should be noted that the checklist shown may be used in the absence of the appliance manufacturer's instructions.

Pre-service Checks

1. Check with the customer to ascertain any problems or faults with the installation.
2. Check the general condition of the appliance and that the installation conforms to appropriate standards.
3. Check that there is adequate ventilation and that the flue is both routed and terminated correctly.
4. Check there are no signs of spillage on the appliance or adjacent walls.
5. Check the operation of all controls, the flame supervision device and ignition system.
6. Check flame picture.
7. Check for any signs of water leakage.
8. Advise the customer of any problems.

Full Service

1. Isolate gas, water and where applicable electricity supplies.
2. Dismantle as necessary and clean dust and deposits from within the casing and/or the fireplace opening.
3. Check for signs of damage to electrical connections, cables or components, clean and rectify as necessary.
4. Remove and clean main burner and injectors.
5. Remove and clean pilot burner and injectors.
6. Where applicable, remove and clean the fan and check any air pressure sensing tubes.
7. Inspect the heat exchanger, flueways and carry out flue flow test.
8. Ease and grease gas taps as necessary.
9. Check all appliance seals and joints.
10. Check all disturbed gas connections.
11. Test the ignition device by lighting the boiler.
12. Check pilot flame and test flame supervision device.
13. Ensure that the working pressure, gas rate and flame picture are correct.
14. Check operation of controls as fitted, including:
    - boiler thermostat
    - room thermostat
    - cylinder thermostat
    - frost thermostat
    - overheat thermostat
    - safety valve.
15. Check system pressure on a sealed system.
16. Check pipework and valves for water leaks.
17. Check ball-valve in feed and expansion cistern.
18. Check operation of all electrical controls.
19. Check for spillage.
20. Leave the appliance in a working order and advise the customer of any further work required.

Cookers

CONTENTS

	Page
INTRODUCTION	1
The Gas Appliances Directive	1
Gas Appliances (Safety) Regulations 1995	1
European Standards	2
Regulations and Standards Affecting Installation and Maintenance	3
Health and Safety at Work etc. Act 1974 (HSW Act)	3
Management of Health and Safety at Work Regulations 1999 (MHSWR)	3
Reporting of Injuries, Diseases and Dangerous Occurrences Regulations 1995 (RIDDOR 95)	3
Building Regulations 2000 and Building (Scotland) Regulations 2004	3
British Standards	4
Manufacturer's Instructions	5
COOKING APPLIANCES	6
Grills	6
Conventional	7
Surface Combustion	8
Additional Information	8
Hobs/Hotplates	8
Spillage Tray	9
Pan Supports	9
Float Rail	10
Hotplate Burners	11
Thermostatically Controlled Hotplate Burners	12
Hotplate and Grill Taps	13
Ovens	14
Thermostats	14
Cooking Temperature Guide	16
Oven Construction	17
Direct Ovens	17
Indirect Ovens	18
Forced Convection	18
Ignition Systems	18
Piezo-electric Crystal	19
Battery or Mains-operated High Voltage Spark Ignition	19
Electrodes	19
Ignition Electrode Leads	19
Flame Supervision Devices	19
Oven Thermostat and Flame Supervision Device Assembly	20
Timers/Clocks	21
Solenoids	21

(continued overleaf)

	Page
INSTALLATION OF COOKING APPLIANCES	22
Preliminary Examination	22
Second-hand Appliances	22
Location	22
Siting	23
Clearance Requirements	23
Ventilation	23
Ventilation Requirements for Cooking Appliances	23
Gas Connections	24
General	24
Connections for 1st, 2nd and 3rd Family Gases	25
Use of an Appliance Flexible Connector	25
Use of Rigid Pipework Connections	26
Stability Devices	26
Electrical Connections	26
COMMISSIONING COOKING APPLIANCES CHECKLIST	30
SERVICING COOKING APPLIANCES	32
Fault Diagnosis on Cooking Appliances	32
Blockages	33
Incorrect Operating Pressure	33
Incorrect Aeration Level	34
Incorrect Assembly of Oven Components	34
Cooker Not Level	34
Cooking Methods Used	34

INTRODUCTION

This introduction contains an appraisal of some of the most important Regulations that determine both the design and manufacture of certain domestic appliances.

To reinforce this, there is reference to the Regulations and British Standards that domestic gas appliance installations must comply with. These Regulations and Standards also impose certain constraints on the subsequent maintenance of appliances and the pipework installation.

THE GAS APPLIANCES DIRECTIVE

Since 1985 the European Union has been pursuing a 'common approach' to tackle the problems associated with trade barriers between member states. For example, EU countries have different laws at present relating to product safety. Removing these barriers is at the heart of the Single European Market introduced in 1992.

The Gas Appliances Directive sets out legal requirements that in future will apply across the European Union. Member countries are required to amend their existing legislation, or to introduce new legislation that conforms with the requirements of the directive. The United Kingdom has implemented the Gas Appliances (Safety) Regulations to conform with the directive.

GAS APPLIANCES (SAFETY) REGULATIONS 1995

Until 1992, the safety (to consumers) of gas appliances sold in the United Kingdom has been covered by the Consumer Protection Act and specifically by the Gas Cooking (Safety) Regulations, and the Heating Appliances (Fire-guards) Regulations. The Gas Appliances (Safety) Regulations 1995 introduced specific requirements.

There was a transitional period until 1996 in which gas appliances offered for sale in the United Kingdom were allowed to meet the old requirements.

The main provision of the new Regulations are:

a) **appliances must be safe**

b) **appliances must be tested**

c) **appliances must be quality guaranteed.**

This means that during the manufacturing process the manufacturer must operate a quality scheme of some type, such as BS 5750, to ensure that all appliances conform to the tested design. This scheme will be monitored by the 'Notified Bodies'.

d) appliances must carry the CE mark

All appliances that conform to provisions **a), b) and c)** will carry a CE mark issued by the 'Notified Bodies' (see Figure 1).

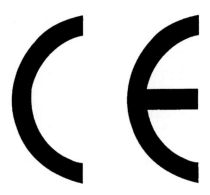

Figure 1 CE mark

The Regulations include detailed procedures for product conformity attestation by third party notified bodies, appointed by the Secretary of State.

All new gas appliances must have information included, covering safe installation, operation and maintenance.

EUROPEAN STANDARDS

European Standards are currently being compiled. For some appliances, where no European Standard is planned, the National Standards (in this country, British Standards) may be recognised. This, for example, will apply to the British type of gas fire.

REGULATIONS AND STANDARDS AFFECTING INSTALLATION AND MAINTENANCE

Health and Safety at Work etc. Act 1974 (HSW Act)

This Act applies to everyone concerned with work activities, ranging from employers, self-employed, and employees, to designers, suppliers and importers of materials for use at work, and people in control of premises. The duties apply both to individual people, and to corporations, companies, partnerships, local authorities etc. Employers have a duty to ensure, so far as is reasonably practicable, the health, safety and welfare at work of all employees, and not to expose people who are not their employees to risks to their health and safety.

Management of Health and Safety at Work Regulations 1999 (MHSWR)

These Regulations impose a duty on employers and self-employed persons to make suitable and sufficient assessment of risks to the health and safety of employees, and non-employees affected by their work. It also requires effective planning and review of protective measures, health surveillance, emergency procedures, information and training.

Reporting of Injuries, Diseases and Dangerous Occurrences Regulations 1995 (RIDDOR 95)

These Regulations require employers to report specified occupational injuries, diseases and dangerous occurrences (events) to the HSE. Certain gas incidents are reportable by suppliers of gas through fixed pipe distribution systems and/or LPG suppliers, and gas installers are required to report certain dangerous gas appliances to the HSE.

Building Regulations 2000 and Building (Scotland) Regulations 2004

These Regulations address the various aspects of building design and construction which include energy conservation and health and safety.

The Secretary of State has approved a number of documents under the Building Regulations 2000 as practical (non-mandatory) guidance to meeting the requirements under the Regulations.

Similar 'deemed to satisfy' guidance is provided in technical handbooks of the Building (Scotland) Regulations 2004.

The documents that particularly relate to gas work in domestic premises are:

- **Building Regulations 2000 (England and Wales)**

    | Part | A | – | Structure |
    | Part | B | – | Fire Safety |
    | Part | F | – | Ventilation |
    | Part | G3 | – | Hot Water Storage |
    | Part | J | – | Combustion Appliances and Fuel Storage Systems |
    | Part | L | – | Conservation of Fuel and Power |
    | Part | M | – | Access To and Use of Buildings |
    | Part | P | – | Electrical Safety |

- **Building (Scotland) Regulations 2004**

    Section 1 – Structure
    Section 2 – Fire
    Section 3 – Environment
    Section 4 – Safety
    Section 6 – Energy

BRITISH STANDARDS

British Standards' specifications are an invaluable guide to the installation of gas appliances. If followed, these standards will satisfy the requirements of current Regulations.

The following is a selection of some of the important British Standards Specifications relating to Domestic Gas Appliances, which give guidance on the minimum standard that appliance installations should comply with, to satisfy current Regulations:

BS 5546: 2000	–	Specification for installation of gas hot water supplies for domestic purposes
BS 5588: (Domestic) Part 1 1990	–	Fire precautions in the design, construction and use of buildings
BS 5864: 2004	–	Installation and maintenance of gas fired ducted warm air heaters of rated input not exceeding 70 kW net (2nd and 3rd family gases)
BS 5871: Part 1 2005	–	Gas fire, convector heaters and fire/back boilers (2nd and 3rd family gases)

BS 5871:	Part 2 2005	–	Inset fuel effect gas fires of a heat input not exceeding 15 kW (2nd and 3rd family gases)
BS 5871:	Part 3 2005	–	Decorative fuel effect gas appliances of a heat input not exceeding 20 kW (2nd and 3rd family gases)
BS 6700:	1997	–	Design, installation, testing and maintenance of water supplies for domestic purposes
BS 8423:	2002	–	Fire-guards for fires and heating appliances for domestic use - Specification
BS 6172:	2004	–	Installation and maintenance of domestic gas cooking appliances (2nd and 3rd family gases) – Specification
BS 6798:	2000	–	Specification for installation of gas fired boilers of rated input not exceeding 70 kW net
BS 6891:	2005	–	Installation of low-pressure gas pipework of up to 35 mm (R 1¼) in domestic premises (2nd family gases)
BS 5440:	Part 1 2000	–	Flues
BS 5440:	Part 2 2000	–	Air supply
BS 5482:	Part 1 2005	–	Code of practice for domestic butane and propane gas burning installations. Installations in permanent buildings, residential park homes and commercial premises up to 28 mm
BS 7624:	2004	–	Installation and maintenance of domestic direct gas fired tumble dryers up to 6 kW heat input (2nd and 3rd family gases) - Specification

MANUFACTURER'S INSTRUCTIONS

Manufacturer's instructions are important for the installation, commissioning, maintenance and use of any gas appliance. These instructions must be read and followed.

After installation of the appliance or subsequent maintenance of it, the instructions must be returned to the consumer so that they may store them for future reference. This includes both user and installation/servicing instructions. (This is a requirement of the Gas Safety Regulations.)

COOKING APPLIANCES

Domestic gas cookers have changed a great deal over the years in their design, construction and performance. Methods of installation have changed, too, especially since the introduction of split-level and built-in cookers in fitted kitchens.

The choice of cooking appliances now offered for sale by manufacturers is comprehensive. The cooker today is not just functional, but must suit the aesthetic needs of customers and complement the modern kitchen.

The consumer is offered a wide choice of colour, size, degree of automation, performance and safety features, plus the benefit of ease of cleaning. The installer requires a good working knowledge of installation, commissioning and fault-finding procedures, along with other technologies, to enable the consumer to gain maximum benefit from their appliance.

Domestic gas cookers normally are made up of three basic components:

- grill
- hob/hotplate
- oven.

There are three classes of appliance:

Class 1 Free-standing, these include double and single oven types, manufactured with high-level or under-hotplate grills

Class 2 Appliances designed to be built into kitchen furniture, these may be either:

a) an appliance made in one complete unit

b) appliances containing separate units, where the hotplate, grill or oven may be installed into different kitchen units or worktops

Class 3 Appliances designed for installing adjacent to kitchen units or worktops (slot-fix)

GRILLS

The grill is mainly used for browning foods or toasting.

The grill can be manufactured and installed to suit one of the following:

- high-level
- below the hotplate
- independent wall-mounted.

There are two main types of grill:

- conventional
- surface combustion.

Conventional

The conventional type of grill consists of a burner, usually made of pressed steel. This burner can be mounted in the middle from front-to-back of the grill, or from side-to-side at the rear of the grill assembly.

The burner is directed so that the flames heat an expanded metal 'fret'. The products of combustion are vented through manufactured holes in the top of the canopy of the grill assembly or, in the case of below-the-hotplate types, through a small outlet built into the top of the appliance.

These types of grill can lead to uneven cooking, due to the burner failing to effectively heat the whole area of the fret.

Burners are normally of simplex design with the primary air level preset by the manufacturer. A simplex burner is the term given to a burner fed by only one injector (see Figure 2).

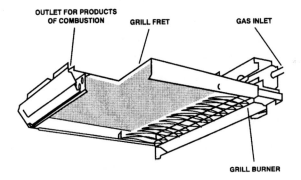

Figure 2 Conventional grill

Surface Combustion

This type of grill is made up of a fine metal mesh or ceramic honeycomb and serves two purposes. It is the grill fret and burner head combined.

Unlike the conventional grill type that takes approximately 50% of its air for combustion through the primary air port, the surface combustion grill entrains 80–90% of combustion air through the primary air port. The whole of the 'fret' becomes the burner, leading to the surface temperature of the fret rising very quickly, with an even distribution of heat across the entire area.

Some burners are manufactured with a steel plate in the middle, preventing the centre of the grill becoming too hot and so distorting the fine mesh (see Figure 3).

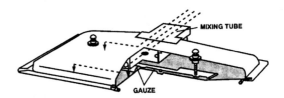

Figure 3 Surface combustion grill

Additional Information

The grill pan is an important part of the design and subsequent performance of any grill. Improvements have been made over the years with the introduction of adjustable height, rotisseries, etc., but it is important to remember that on most high-level grills, the grill pan must be located correctly to prevent the products of combustion, from the rear hotplate burners, causing 'vitiation' on the grill burner.

(Vitiation is a term used when products of combustion are entrained as air for combustion, thus contaminating the combustion air with carbon dioxide and reducing its oxygen content. This may lead to incomplete combustion and flame lengthening.)

HOBS/HOTPLATES

The hob or hotplate is used for boiling, simmering, frying, pot-roasting, braising and steaming. It can be manufactured either as part of a free-standing or built-in cooker, when it is usually referred to as a hotplate, or as a separate unit installed in a worktop, when it is usually called a hob.

The burners, burner mixing tubes, ignition leads and electrodes may be part of the hotplate assembly, or they may be covered by the hotplate top. A hotplate top may be loose or secured by screws.

The majority of cookers on the UK market have four hotplate burners. The burners may or may not be of an identical size and rated heat inputs (see Figure 4).

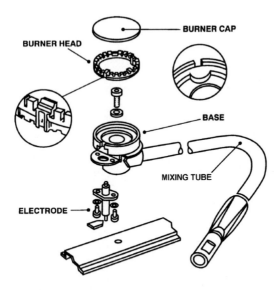

Figure 4 A typical hotplate assembly

Spillage Tray

In the event of a pot boiling over, a suitable catchment area for the liquid is essential. The design of the hotplate assembly allows for spillage of liquids to be contained, either in a well formed in the hotplate top, or by an individual, removable spillage tray around each burner.

Cookers are manufactured to the requirements of BS EN 30, which give specific sizes of spillage trays for liquid catchment. These help to prevent the ingress of liquids below the hotplate, and also prevent any spillage over the edge of the hotplate.

Pan Supports

Pan supports are designed to provide adequate support for any normal cooking pan. They have been made in various materials over the years, including vitreous enamel coated cast iron, aluminium and stainless steel. They may be individual to a burner or cover two or more burners.

Pan supports are also designed to allow for the fact that the burner will require secondary air to maintain proper combustion. A further consideration is their ability to allow the transfer of heat from the burner to the pan, without the flame impinging on the supports.

It is important to remember that pan supports are dedicated to the appliance they were designed for and are not normally interchangeable.

Float Rail

This is the term given to the manifold that distributes gas to the hotplate burners via individual taps. Normally the grill and oven taps are also fitted to the float rail.

The float rail is located directly behind the hotplate tap knobs. It can be manufactured from various materials, including:

- cast iron

- alloy or steel.

It can include mounting points for additional equipment, including:

- ignition micro-switches

- oven light switches.

The gas supply is generally connected to the float rail from the rear of the cooker. Most cookers manufactured in the UK with 'drop down' hotplate lids are protected by a safety control that shuts off the gas to the hotplate taps when the lid is closed.

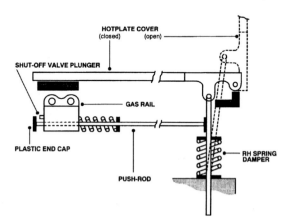

Figure 5 Hotplate lid safety control

Hotplate Burners

Hotplate burners are designed so that the flames from the burner head impinge on the cooking utensil and maintain a good combustion process over varying gas rates, from full-on to simmer.

Burners are designed and manufactured in a variety of materials and finishes, including:

- cast iron
- light alloy
- pressed steel.

Burner caps on modern appliances generally are removable, to assist the consumer in cleaning. However, there is a trend towards manufacturing hotplate burners as part of the hotplate tap using gasket seals to maintain gas tightness. Any work on these types of hotplate requires careful attention to the gaskets.

Modern cookers have fixed primary air ports, but on older cookers these air ports were normally adjustable by slides or shutters. Examples of these are given in the following diagram and should be adjusted to give satisfactory combustion (Figure 6).

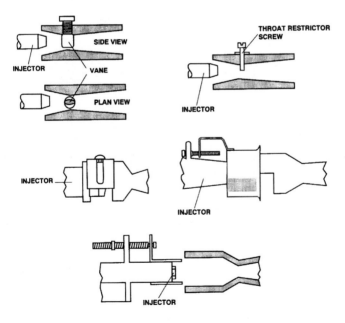

Figure 6 Typical examples of primary air adjustment

An important design feature of any burner is flame retention. On a cooker this is normally achieved either by a series of smaller flame ports under the main burner ports, or by flame retention rings fitted underneath the burner caps. A proportion of the gas/air mixture that enters the burner will be 'slowed down' to enable the main flame to be stabilised, thereby providing continuous ignition to the flames burning from the main burner ports, as indicated in the following diagram (Figure 7).

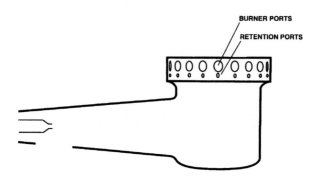

Figure 7 Typical hotplate burner

If liquid boils over onto the burner the flame retention ring may become blocked leading to 'flame lift' from the main burner ports.

Thermostatically Controlled Hotplate Burners

Some hotplates incorporate a single burner with a thermostatic control. This type of hotplate is controlled via the hotplate tap, which includes a bellows type thermostat. A contact sensor connected to a capillary tube is mounted within the burner head. This sensor is in contact with the cooking vessel and will adjust the gas rate to the burner accordingly, dependent on the thermostat setting made by the consumer (see Figure 8).

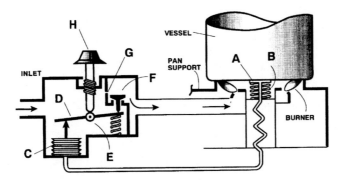

Liquid thermostat for control of cooker hotplate burner. A disc-shaped thermal element, **A**, is kept in contact with the cooking vessel by spring, **B**. Movement of bellows, **C**, is transmitted by rod, **D**, pivoted at **E**, to operate gas valve, **F**. **G** is a by-pass, and **H**, a thermostat setting knob to select different cooking heats.

Figure 8 Thermostatically controlled hotplate burner

Hotplate and Grill Taps

Most cooker hotplates and grills use a plug and taper tap. The tap consists of a taper plug and body. The tap and body are engineered to produce a close fit. These are usually machined from alloy or brass and the surfaces between the two are lubricated with high temperature grease. A niting plate is used to hold the two together and this also dictates the movement and action of the tap.

As the taps are being subjected to a hot working environment, the grease, with use, may require replacing (a suitable grease should be used such as molycote). The main symptoms of a lack of grease are stiff operation or the tap letting-by. These situations will need immediate rectification (see Figure 9).

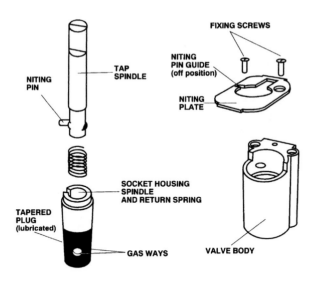

Figure 9 Typical hotplate tap

Another type of tap found on hotplates and grills is a Labyrinth tap. These taps have a 'labyrinth' disc machined with many routes for the gas to pass along, depending on the setting. These taps were introduced to give better simmer control.

As the disc turns the flow of gas is restricted or increased as the slots and holes align.

OVENS

In the domestic kitchen the oven is mainly used for baking, roasting, braising, or plate and food warming.

Thermostats

Most gas ovens are controlled by a thermostat graduated in 'Gas Marks', as opposed to the temperature settings on electric ovens. This form of graduated temperature control is unique to UK cookers and provides a general guide to cooking temperature between appliances supplied by one cooker manufacturer and another.

Most modern thermostats directly control the flow of gas to the oven burner. They contain a liquid which when heated acts upon bellows, which in turn operate the gas valve. A by-pass allows sufficient gas supply to the oven burner to maintain temperature and keep the burner alight (Figure 10).

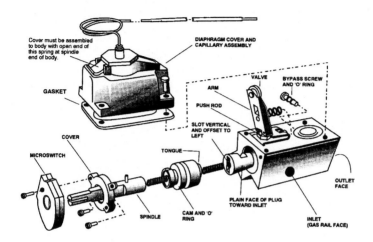

Figure 10 Bellows type oven thermostat

Older cookers were often manufactured with a rod type thermostat as in Figure 11.

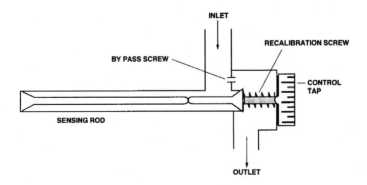

Figure 11 Rod type oven thermostat

Cooking Temperature Guide

Cooking process	Oven heat	Gas mark	Electric oven temperature °C	°F
Plate warming	very cool	¼	110	225
Keeping food hot, milk puddings	very cool	½	120	250
Egg custards	cool	1	140	275
Rich fruit cakes, braising	cool	2	150	300
Low temperature roasting, shortbread	moderate	3	160	325
Victoria sandwich, plain fruit cake, baked fish	moderate	4	180	350
Small cakes, choux pastry	fairly hot	5	190	375
Short pastry, Swiss rolls, soufflés	fairly hot	6	200	400
High temperature roasting, flaky pastry, scones	hot	7	220	425
Puff pastry, bread	very hot	8	230	450
Small puff pastries, browning cooked foods	very hot	9	240	475

These temperatures relate to the centre oven temperature.

Oven Construction

The oven is basically an insulated metal box, with an access door. The oven contains an opening at the bottom near to the burner, for the supply of combustion air, and an outlet at the top for combustion products.

The modern oven has a temperature control, a method of flame supervision, an ignition system, possibly an automatic time control, and some have an electric light.

The door of the oven may be either 'side opening' or 'drop down' and may be made of (or contain) a heat-resistant glass viewing panel, to enable the consumer to look at the cooking process without opening the door. The door will close against a heat resistant seal. This seal may either be part of the door assembly or fitted around the opening of the oven's main frame.

For an oven to perform correctly the door seal needs to be effective.

Many ovens have, as an option, 'assist clean' linings. These are covered with a special coating that resists stains and has the ability to burn off fatty substances. Two oven shelves are normally provided which can be positioned at different heights.

Direct Ovens

Most cookers manufactured in the UK have a direct oven. The combustion process is within the oven, therefore leading to natural convection. This allows for different heat zones within the oven and the cooking of various dishes at the same time, on different shelves of the oven (hot at the top, cooler lower down), (see Figure 12).

Figure 12 Direct oven

Indirect Ovens

American and Continental manufactured ovens tend to be indirect. The combustion process is separated from the cooking process. The main advantage of this type of oven is a more even oven temperature distribution (Figure 13).

Forced Convection

To increase the efficiency and to maintain an even temperature, forced circulation is sometimes employed. This increases the dependency on electricity, using a fan to distribute the heat, as shown in the diagram below.

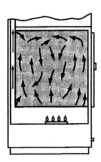

Figure 13 Indirect oven

IGNITION SYSTEMS

The cooker ignition system has developed over the years. When cookers were first introduced around 150 years ago, lighting was a manual task with match or taper. During the late 1950s the pilot light was introduced to provide ignition. This has been superseded by electric spark ignition, introduced in the early 1970s, which has become the standard for gas cookers.

Modern cookers employ high voltage electric spark ignition systems to ignite the burners. This may be achieved by either:

- piezo-electric crystal
- battery- or mains-operated spark ignition.

Piezo-electric Crystal

This is a crystal that when mechanically compressed, or when the compression is released, will produce a high electrical voltage. This high voltage is directed through a special insulated wire to an electrode. There is enough electrical potential to cause a spark to occur between the electrode tip and the burner. This particular method is generally used on budget-priced cookers.

Battery or Mains-operated High Voltage Spark Ignition

The most common ignition systems used on gas cookers fall into this category. The power is supplied either from a 1.5 volt battery or from the household mains. These methods use a capacitor to build up the charge, and a trigger device within the electronic circuit of the spark generator allows a voltage of between 10,000 and 15,000 volts to be passed to an electrode positioned in the gas stream, the resulting spark igniting the gas.

High voltage ignition systems may also employ flame conductance or flame rectification for flame supervision purposes.

Electrodes

An ignition electrode is similar to a spark plug in a car, although physically smaller. It is an essential part of a high voltage spark ignition system. The gap between the electrode tip and the earth point has to be correct, usually 3–5 mm.

Electrodes are encased in a ceramic material and need to be treated with care. They vary in shape and size, are manufactured for specific appliances and are rarely interchangeable.

Ignition Electrode Leads

Ignition leads are an important part of a cooker's ignition system and care should be taken when replacing, as leads trapped between metallic surfaces may result in the spark 'shorting' to earth.

FLAME SUPERVISION DEVICES

Since 1972, flame supervision devices have been a requirement of British Standards on all cooker ovens with an input rating in excess of 600 W (2,000 Btu/h).

The most common type is shown in Figure 14. The shut-down time is the period taken for the device to fail safe, by shutting off the gas supply to the oven. It is the responsibility of any person working on these appliances to ensure that flame supervision devices conform to the relevant standard for shut-down.

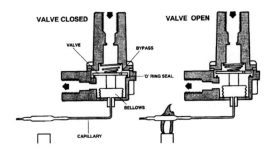

Figure 14a Flame supervision devices

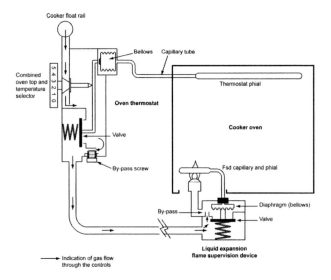

Figure 14b Oven thermostat and flame supervision device assembly

OVEN THERMOSTAT AND FLAME SUPERVISION DEVICE ASSEMBLY

The interaction of these controls when fitted in series is:

- when put into operation when the oven is cold, full gas rate will flow through the thermostat, but only on by-pass rate through the flame supervision device

- within a short period (30 – 60 seconds) the FSD sensing phial will have heated up causing the expansion of the liquid within it to open the gas valve fully in the FSD, thus allowing full gas rate to the burner

- the temperature of the oven continues to be monitored by the thermostat sensing phial (located at the top of the oven) and when satisfied reduces the gas flow to its by-pass rate

- this by-pass rate is sufficient to maintain the FSD sensing phial hot enough to keep its valve fully open.

Faults

If either of the thermostat or FSD by-pass screws become blocked then the burner will go out. To identify which screw is at fault:

- no gas to the burner when the oven is cold when first lit will normally indicate it is the FSD by-pass that is blocked;

- no gas to the burner when the thermostat is satisfied and the oven is hot indicates it is the thermostat by-pass that is blocked.

TIMERS/CLOCKS

These may either be analogue or digital. Their main function is to provide a facility for automatic pre-timed cooking in the oven. The manufacturer's instructions to the user give guidance on setting the timer/clock. It is important, however, to ensure that the consumer is aware that the timer/clock will need setting back to the manual position after automatic cooking (Figure 15 shows a typical analogue oven timer).

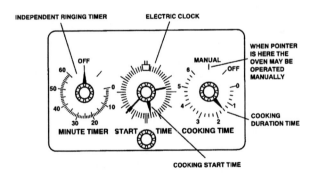

Figure 15 Analogue oven timer

SOLENOIDS

Some cookers include a solenoid to control the gas to the oven burner. The oven timer/clock energises the solenoid initially. The solenoid will then be controlled by the oven micro-switch and linked electrically through the timer/clock.

INSTALLATION OF COOKING APPLIANCES

The following considerations must be taken into account when installing a cooking appliance (cooking appliance means either a cooker, or a hotplate, oven, grill or griddle, or any combination of these in a single unit).

PRELIMINARY EXAMINATION

The appliance data badge shall be examined to ensure that the gas supplied and the operating pressure are suitable:

    i.e. natural gas   G20

    propane   G31

    butane   G30

If the data plate is not readable the appliance should not be installed.

SECOND-HAND APPLIANCES

When a second-hand appliance is to be installed, or when an appliance is moved from one location to another, the physical condition of the appliance should be checked to ensure that it is free from deterioration, distortion or displacement of any components that may adversely affect its safety in use. If the manufacturer's instructions are not available they should be obtained from the supplier (manufacturer or agent) before the appliance is installed.

LOCATION

The appliance shall be installed in accordance with the manufacturer's instructions, and shall not be installed in a bathroom or shower room.

Unless of single hotplate burner type, the appliance shall not be installed in a bed-sitting room of volume less than 20 m^3.

Third family (propane and butane) gas appliances shall not be installed in a room or internal space below ground level e.g. in a basement or cellar. This does not prevent the installation where the room or space are basements with respect to one side of the building, but which are open to ground level on the opposite side.

Siting

The manufacturer's instructions shall be complied with, and consideration should be given to:

- convenience in relation to other facilities such as work surfaces, sinks and other appliances
- customer requirements
- draughts, which should be avoided by siting away from doors or openable windows
- restrictions to the use of doors, kitchen furniture or utensils should be avoided
- avoid siting close to curtains or other combustible furnishings
- the effect on adjacent fridge/freezers
- carpets, which should be suitable and secured.

Clearance Requirements

In the absence of specific manufacturer's instructions, Figure 16 gives general guidance on clearance zones for combustible materials, which should be followed.

A floor-standing appliance shall be sited on a level, stable base, and if it is installed by means of an appliance flexible connector, an adequate level surface in front of the appliance shall be provided to allow it to be moved forward enough for disconnection.

VENTILATION

The total ventilation requirements for the appliance and other appliances installed in the same room or location shall comply with BS 5440-2, which is explained in the ConstructionSkills publication *Gas Safety – Ventilation Requirements*.

The effect of cooker hoods and other extractors (extractor fans, tumble dryers) on additional gas appliances installed in the same room or location, or in adjacent rooms, must be taken into consideration.

Any cooker hood manufacturer's instructions take precedence over the cooker manufacturer's instructions.

Ventilation Requirements for Cooking Appliances

An appliance means: domestic oven, grill, griddle, hotplate (two or more burners) or any combination thereof.

A window opening direct to outside (or equivalent) must always be provided in the room or space in which the appliance is installed.

Alternative acceptable forms of opening that are deemed equivalent to an opening window are:

- an adjustable louvre
- a hinged panel
- any other means of ventilation that opens directly to outside.

The free area of a purpose provided vent when a cooking appliance is installed in rooms must take into account:

- the volume of a room
- whether a door opening direct to outside from the room has an effect on this required vent.

The table below lists these requirements:

Volume of room (m^3)	Is there an opening door direct to outside?	Free area of purpose provided vent (cm^2)
less than 5	Not applicable	100
more than 5 but less than 10	No	50
more than 5 but less than 10	Yes	Nil
more than 10	Not applicable	Nil

GAS CONNECTIONS

General

The installation pipework and final connection to the appliance shall be sized sufficiently to maintain the heat input of the appliance as specified by the manufacturer.

Where an existing appliance is to be replaced by another, it is good practice to check before disconnecting that its operating pressure under maximum gas flow conditions is satisfactory.

A visual inspection should be made of any existing pipework to ascertain its suitability. If breaking into the gas supply pipe is necessary, then a test for gas tightness should be completed before any work commences. The location of the appliance termination point shall be in accordance with the manufacturer's instructions.

Connections for 1st, 2nd and 3rd Family Gases

For 2nd family gases (natural gas) the installation pipework to the termination point shall be in accordance with BS 6891, as detailed in the ConstructionSkills publication *Gas Safety*.

For 1st family gases (manufactured gas and LPG/air), because there is no current standard, due regard should be taken of BS 6891.

For 3rd family gases (propane and butane) the installation pipework to the termination point shall be in accordance with BS 5482-1, 2 or 3 as appropriate.

The appliance connection shall be by means of a suitable appliance flexible connector for use with a self sealing plug-in bayonet device (complying with BS 669), or if necessary, by rigid pipework.

Use of an Appliance Flexible Connector

An appliance flexible connector should not be:

- subjected to undue force while either connecting or disconnecting

- subjected to excessive heat by direct exposure to flue products or by contact with any hot surface

  - if the appliance has not been certified by a third party as being in accordance with BS 5386-3 or 4 or EN 30, or is NOT labelled to indicate the surface temperature of the appliance, it should be connected by rigid pipework

  - appliances badged to the above standard which give a temperature rise greater than 70°C, will have a note to this effect in the installation instructions

- positioned where it will suffer mechanical damage, e.g. abrasion from the surrounding kitchen furniture which may be moved in use such as a drawer or door, or by being trapped by any stability device etc.

- used on 3rd family gases unless indicated by a continuous red marking along its length.

The socket into which the plug of the flexible connector fits should be permanently attached to a firmly fixed gas installation pipe and positioned such that:

- the appliance flexible connector hangs freely down (see Figure 17)

- the plug-in bayonet connector should be accessible for disconnection after moving the appliance.

Use of Rigid Pipework Connections

Where rigid pipework is used, an isolation device (tap) with means of disconnection should be provided wherever practicable.

Stability Devices

A free-standing domestic cooker, using a flexible connector, shall be fitted with a stability device firmly secured to the fabric of the building.

Where the appliance manufacturers do not provide a purpose designed bracket described in the installation instructions, then suitable devices as shown in Figure 18 shall be used.

Electrical Connections

The electrical connections to the appliance shall comply with the Regulations for Electrical Installations, and be in accordance with any manufacturer's instructions with regard to connection method, fuse rating, earth connection and voltage range.

The connection to the mains electricity shall be readily accessible and not more than 1.5 m to the appliance, and be capable of electrical isolation.

To encourage the removal of the electrical plug whenever work is carried out on the appliance, an unswitched socket is recommended. Where a switched socket is used, the installer should satisfy himself that the electrical polarity is correct.

The flexible lead and the plug and socket shall not be directly exposed to flue products and shall not be in contact with hot surfaces.

Sockets should only be sited outside the clearance zones indicated in Figure 16.

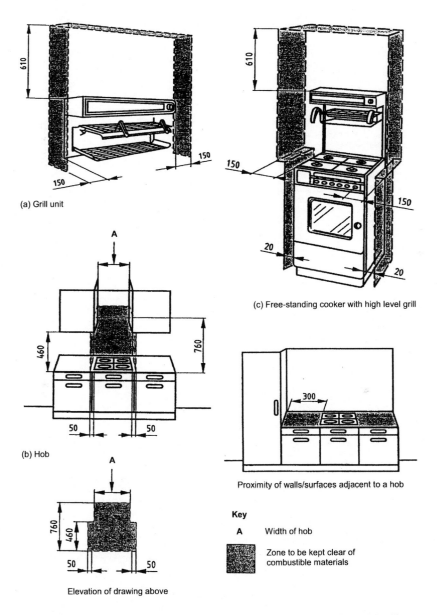

Figure 16 Proximity of combustible materials to cookers, hotplates and grills

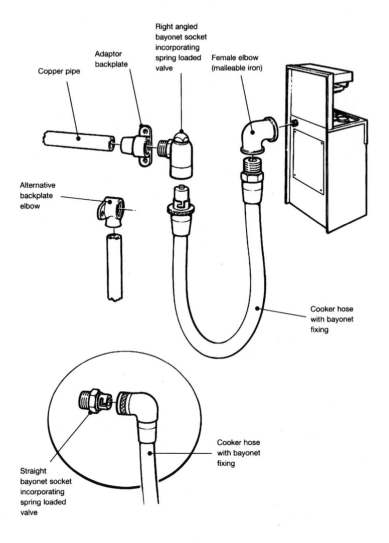

Figure 17 Typical appliance flexible connectors for domestic cookers for 1st, 2nd and 3rd family gases

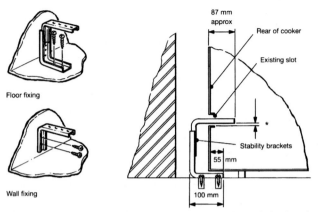

(a) Cooker stability bracket (for cookers specifically designed with bracket engagement slot)

* The bracket to be adjusted to give the smallest practical clearance between the bracket and the bottom of the engagement slot in the rear of the cooker

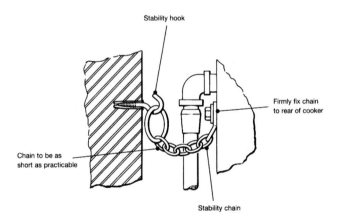

(b) Cooker stability chain (for cookers not designed with bracket engagement slot)

Figure 18 Methods for provision of cooker stability

COMMISSIONING COOKING APPLIANCES CHECKLIST

The appliance installation shall be checked for tightness and purged. Where an appliance has a drop-down lid on the hob, THIS MUST BE IN THE RAISED POSITION when completing a tightness test to ensure that any lid safety cut-off device is not isolating the gas supply to the hob.

The appliance operating (working) pressure shall be checked in accordance with the manufacturer's instructions.

In the absence of manufacturer's instructions, then the following procedure should be used:

- for a free standing cooker check the operating pressure with three hotplate burners on full

- for any other appliance, with the burners on full (not exceeding three burners in total).

The appliance shall be operated to ensure that all burners and controls function correctly as detailed in the manufacturer's instructions.

The customer should be instructed on the correct use and operation of all user controls, and be handed ALL INSTRUCTIONS issued with the appliance.

When commissioning appliances for which customers no longer have the installation and user instructions, or the instructions are inadequate, then the following procedure for commissioning may be used.

After connecting the appliance and testing its connections for tightness in accordance with foregoing procedures, then the following checks should include that:

- the appliance has been correctly assembled

- all air has been satisfactorily purged

- the appliance operating (working) pressure is satisfactory

- all ignition devices operate satisfactorily

- all flame supervision devices operate safely

- any lid safety cut-off device operates correctly

- all flame pictures are satisfactory

    – hotplate burners are well aerated

    – grill and oven burners may have slacker flames but without yellow tips

- the thermostat by-pass rate is satisfactory, by heating up the oven and then turning down to a low setting

- all other devices (automatic cooking control timers, lights, rotisseries etc.) operate correctly

- the oven door seal is satisfactory by checking its grip on a 50 mm strip of paper shut in the door

- a satisfactory stability device has been fitted, and the appliance is level (front to back and side to side)

- the supply of combustion air is satisfactory

- the safe operation of the appliance has been explained to the customer.

SERVICING COOKING APPLIANCES

Most cooking appliances are kept clean by the customers on a regular basis, but there is still a need for periodic servicing. The operations should include the commissioning procedure outlined above, and rectifying any problems that are encountered by:

- easing and greasing any stiff burner control taps
- clearing or replacing faulty pilot jets or injectors
- cleaning burner, retention and cross-lighting ports
- cleaning burner mixing tubes and flash tubes
- replacing faulty spark or reignition probes
- readjusting primary aeration devices as necessary
- checking correct location of thermostat and FSD phials
- checking that oven flueways are clear
- checking all electrical connections/connectors are undamaged and secure
- checking electric supply earthed and correctly connected (polarity etc.)
- replacing batteries as appropriate.

FAULT DIAGNOSIS ON COOKING APPLIANCES

Faults which may occur on control devices are detailed in the ConstructionSkills publication *Gas Safety*.

Other common faults are due to:

- blockages
- incorrect operating pressure
- incorrect aeration level
- incorrect assembly of oven components
- cooker not level
- cooking methods used.

Blockages

Blockages of jets and burners may be due to:

- cooking spillages and vapours
- fats and oils splashing and then burnt on
- corrosion
- moisture.

These can cause:

- reduction in primary aeration
- ignition problems with sparks or flash tubes
- FSDs failing to safety
- blockage of oven flues causing flame extinguishment due to reduction of combustion air (vitiation)
- uneven heating on grills due to blocked burner ports.

Incorrect Operating Pressure

This will be either due to:

- incorrect meter regulator setting
- incorrect installation pipework, from either:
    a) pipe sizes too small, or
    b) partially blocked supplies.

The cooking appliance operating pressure should be checked when:

- only three hotplate burners on full, with all other appliances OFF
- all cooker burners on full with all other appliances on FULL.

The operating pressure at the meter test nipple should read 21 ± 1 mbar.

The operating pressure at the appliance should be within 1 mbar of the meter regulator pressure. High pressure can cause unstable flames. Low pressure can cause heat input reduction, poor performance and flame extinguishment on simmer or by-pass settings.

Incorrect Aeration Level

Over-aeration can cause unstable flames and ignition problems. Under-aeration can cause poor flame pictures, yellow tipping, smells from hotplate burners and ignition problems, especially from flash tubes and spark ignition systems.

Incorrect Assembly of Oven Components

This can cause uneven cooking due to:

- misplaced oven linings
- shelves not correctly positioned
- shelves distorted.

Cooker Not Level

The main cooking complaint would be uneven cooking in the oven.

Cooking Methods Used

These will include:

- incorrect size of tins or baking trays used
- incorrect shelf height used
- incorrect thermostat setting used
- incorrect time allowed.

Many of these complaints arise from the customer using a replacement appliance and not following the user instructions provided.

If the oven appears to be operating normally, and the cooking method appears satisfactory, the problem may be due to a faulty thermostat. Thermostats are usually designed for replacement rather than recalibration.

Ducted Air Heaters

CONTENTS

	Page
INTRODUCTION	1
The Gas Appliances Directive	1
Gas Appliances (Safety) Regulations 1995	1
European Standards	2
Regulations and Standards Affecting Installation and Maintenance	3
Health and Safety at Work etc. Act 1974 (HSW Act)	3
Management of Health and Safety at Work Regulations 1999 (MHSWR)	3
Reporting of Injuries, Diseases and Dangerous Occurrences Regulations 1995 (RIDDOR 95)	3
Building Regulations 2000 and Building (Scotland) Regulations 2004	3
British Standards	4
Manufacturer's Instructions	5
DUCTED AIR CENTRAL HEATING	6
Types of Ducted Air Heaters	6
Upflow	6
Downflow	7
Horizontal or Wall-hung	7
Basement	7
Installation Methods	8
Flue Positions	9
Building Heat Loss and Comfort Conditions	10
Ducts	10
Types of Duct System	11
Stub Duct System	11
Fully Ducted System	11
Return Air Paths	13
Open Flued Appliances	13
Positive Return Air	14
Combustion Ventilation	15
Compartment Ventilation	15
Open Flued	15
Room Sealed	16
Ducted Air Heater Controls	16
Flame Supervision Devices	16
Thermo-electric	16
Oxygen Depletion System (ODS) – Atmosphere Sensing Device (ASD)	19
Flue Safety Devices	22
Flame Rectification	23

(continued overleaf)

	Page
Fused Power Supply	25
Timing Controls (Clocks and Timers)	25
Room Thermostat	26
Limit Thermostat	26
Fan Thermostat	26
Fan	27
Transformer	27
Wiring	27
Modairflow	29
Control of the Modairflow System	29
INSTALLATION OF DUCTED AIR HEATERS	**31**
Location and Installation Guidelines	31
Flues and Ventilation	32
Bathrooms and Shower Rooms	32
Living Rooms, Kitchens, Utility Rooms, Halls and Passageways	32
Compartments and Cupboards	33
Airing Cupboards	33
Under Stairs Cupboards	34
Slot-fit Installations	34
Roof Space Installations	34
COMMISSIONING CHECKLIST FOR DUCTED AIR HEATERS	**35**
MAINTENANCE CHECKLIST FOR DUCTED AIR HEATERS	**37**
Pre-service Checks	37
Full Service	37
FAULT-FINDING CHECKLIST FOR DUCTED AIR HEATERS	**41**

INTRODUCTION

This introduction contains an appraisal of some of the most important Regulations that determine both the design and manufacture of certain domestic appliances.

To reinforce this, there is reference to the Regulations and British Standards that domestic gas appliance installations must comply with. These Regulations and Standards also impose certain constraints on the subsequent maintenance of appliances and the pipework installation.

THE GAS APPLIANCES DIRECTIVE

Since 1985 the European Union has been pursuing a 'common approach' to tackle the problems associated with trade barriers between member states. For example, EU countries have different laws at present relating to product safety. Removing these barriers is at the heart of the Single European Market introduced in 1992.

The Gas Appliances Directive sets out legal requirements that in future will apply across the European Union. Member countries are required to amend their existing legislation, or to introduce new legislation that conforms with the requirements of the directive. The United Kingdom has implemented the Gas Appliances (Safety) Regulations to conform with the directive.

GAS APPLIANCES (SAFETY) REGULATIONS 1995

Until 1992, the safety (to consumers) of gas appliances sold in the United Kingdom has been covered by the Consumer Protection Act and specifically by the Gas Cooking (Safety) Regulations, and the Heating Appliances (Fire-guards) Regulations. The Gas Appliances (Safety) Regulations 1995 introduced specific requirements.

There was a transitional period until 1996 in which gas appliances offered for sale in the United Kingdom were allowed to meet the old requirements.

The main provision of the new Regulations are:

a) **appliances must be safe**

b) **appliances must be tested**

c) **appliances must be quality guaranteed.**

This means that during the manufacturing process the manufacturer must operate a quality scheme of some type, such as BS 5750, to ensure that all appliances conform to the tested design. This scheme will be monitored by the 'Notified Bodies'.

d) appliances must carry the CE mark

All appliances that conform to provisions **a), b) and c)** will carry a CE mark issued by the 'Notified Bodies' (see Figure 1).

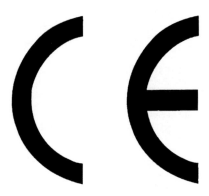

Figure 1 CE mark

The Regulations include detailed procedures for product conformity attestation by third party notified bodies, appointed by the Secretary of State.

All new gas appliances must have information included, covering safe installation, operation and maintenance.

EUROPEAN STANDARDS

European Standards are currently being compiled. For some appliances, where no European Standard is planned, the National Standards (in this country, British Standards) may be recognised. This, for example, will apply to the British type of gas fire.

REGULATIONS AND STANDARDS AFFECTING INSTALLATION AND MAINTENANCE

Health and Safety at Work etc. Act 1974 (HSW Act)

This Act applies to everyone concerned with work activities, ranging from employers, self-employed, and employees, to designers, suppliers and importers of materials for use at work, and people in control of premises. The duties apply both to individual people, and to corporations, companies, partnerships, local authorities etc. Employers have a duty to ensure, so far as is reasonably practicable, the health, safety and welfare at work of all employees, and not to expose people who are not their employees to risks to their health and safety.

Management of Health and Safety at Work Regulations 1999 (MHSWR)

These Regulations impose a duty on employers and self-employed persons to make suitable and sufficient assessment of risks to the health and safety of employees, and non-employees affected by their work. It also requires effective planning and review of protective measures, health surveillance, emergency procedures, information and training.

Reporting of Injuries, Diseases and Dangerous Occurrences Regulations 1995 (RIDDOR 95)

These Regulations require employers to report specified occupational injuries, diseases and dangerous occurrences (events) to the HSE. Certain gas incidents are reportable by suppliers of gas through fixed pipe distribution systems and/or LPG suppliers, and gas installers are required to report certain dangerous gas appliances to the HSE.

Building Regulations 2000 and Building (Scotland) Regulations 2004

These Regulations address the various aspects of building design and construction which include energy conservation and health and safety.

The Secretary of State has approved a number of documents under the Building Regulations 2000 as practical (non-mandatory) guidance to meeting the requirements under the Regulations.

Similar 'deemed to satisfy' guidance is provided in technical handbooks of the Building (Scotland) Regulations 2004.

The documents that particularly relate to gas work in domestic premises are:

- **Building Regulations 2000 (England and Wales)**

    Part A – Structure

    Part B – Fire Safety

    Part F – Ventilation

    Part G3 – Hot Water Storage

    Part J – Combustion Appliances and Fuel Storage Systems

    Part L – Conservation of Fuel and Power

    Part M – Access To and Use of Buildings

    Part P – Electrical Safety

- **Building (Scotland) Regulations 2004**

    Section 1 – Structure

    Section 2 – Fire

    Section 3 – Environment

    Section 4 – Safety

    Section 6 – Energy

BRITISH STANDARDS

British Standards' specifications are an invaluable guide to the installation of gas appliances. If followed, these standards will satisfy the requirements of current Regulations.

The following is a selection of some of the important British Standards Specifications relating to Domestic Gas Appliances, which give guidance on the minimum standard that appliance installations should comply with, to satisfy current Regulations:

BS 5546: 2000	–	Specification for installation of gas hot water supplies for domestic purposes
BS 5588: (Domestic) Part 1 1990	–	Fire precautions in the design, construction and use of buildings
BS 5864: 2004	–	Installation and maintenance of gas fired ducted warm air heaters of rated input not exceeding 70 kW net (2nd and 3rd family gases)
BS 5871: Part 1 2005	–	Gas fire, convector heaters and fire/back boilers (2nd and 3rd family gases)

BS 5871:	Part 2 2005	–	Inset fuel effect gas fires of a heat input not exceeding 15 kW (2nd and 3rd family gases)
BS 5871:	Part 3 2005	–	Decorative fuel effect gas appliances of a heat input not exceeding 20 kW (2nd and 3rd family gases)
BS 6700:	1997	–	Design, installation, testing and maintenance of water supplies for domestic purposes
BS 8423:	2002	–	Fire-guards for fires and heating appliances for domestic use - Specification
BS 6172:	2004	–	Installation and maintenance of domestic gas cooking appliances (2nd and 3rd family gases) – Specification
BS 6798:	2000	–	Specification for installation of gas fired boilers of rated input not exceeding 70 kW net
BS 6891:	2005	–	Installation of low-pressure gas pipework of up to 35 mm (R 1¼) in domestic premises (2nd family gases)
BS 5440:	Part 1 2000	–	Flues
BS 5440:	Part 2 2000	–	Air supply
BS 5482:	Part 1 2005	–	Code of practice for domestic butane and propane gas burning installations. Installations in permanent buildings, residential park homes and commercial premises up to 28 mm
BS 7624:	2004	–	Installation and maintenance of domestic direct gas fired tumble dryers up to 6 kW heat input (2nd and 3rd family gases) - Specification

MANUFACTURER'S INSTRUCTIONS

Manufacturer's instructions are important for the installation, commissioning, maintenance and use of any gas appliance. These instructions must be read and followed.

After installation of the appliance or subsequent maintenance of it, the instructions must be returned to the consumer so that they may store them for future reference. This includes both user and installation/servicing instructions. (This is a requirement of the Gas Safety Regulations.)

DUCTED AIR CENTRAL HEATING

Ducted air central heating differs from conventional central heating by using air as the heating medium instead of water. Air passes across a heat exchanger and the heated air is circulated throughout the property via ducting. The cool air is returned to the appliance via a separate duct to be heated and circulated again.

The installation of a ducted air central heating system is a completely different science from a wet central heating system. Installers involved in fitting and servicing these appliances and systems, require different specialist knowledge and training.

Ducted air heating is usually considered as an alternative to other central heating systems at the time the property is being built. This enables the duct routes to be planned during the property's design and build phases. Where an existing ducted air heater requires exchanging, it is logical to consider direct replacement with another ducted air heater.

All ducted air heaters must be installed to conform to the requirements of both Gas Safety Regulations and Building Regulations. They must be installed to the manufacturer's instructions and should comply with British Standards that give guidance on the minimum standards for these appliances, i.e. BS 5864: Installation of gas-fired ducted warm air units.

This reference manual does not cover this subject in its entirety; however, it does cover the most important points.

TYPES OF DUCTED AIR HEATERS

There are four basic types of ducted air heaters which are generally described by the direction of the air flow through them.

Upflow

Cool air is drawn by a fan into the heater at the bottom, is driven over the heat exchanger in an upward direction and the warmed air leaves the heater at the top.

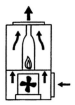

Figure 2 Upflow

Downflow

Cool air is drawn by a fan into the heater at the top and is driven over the heat exchanger in a downward direction. A fan blows the warm air down into the plenum chamber and out through the ducts.

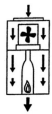

Figure 3 Downflow

Horizontal or Wall-hung

Circulating fan and heat exchanger are situated side by side. The air is driven through the heat exchanger horizontally to the duct.

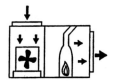

Figure 4 Horizontal or wall-hung

Basement

The air inlet and outlet are both situated at the top of the heater. Cool air is drawn by the fan into the top of the heater, travels across the heat exchanger and is driven up into the dwelling via the warm air outlet, plenum box and ducting.

Figure 5 Basement

INSTALLATION METHODS

The majority of ducted air heaters are installed in cupboards or compartments where their physical size is not critical.

In addition there are two other models available. The storey height model is designed to fit into a corner of a room, the appliance case extending from floor to ceiling enclosing the flue and return air system.

The slot fix model is designed to slot into an opening with just the front of the appliance on show.

a) Compartment

b) Storey height

c).Slot fix

Figure 6 Methods of installation

Flue Positions

To enable flexibility in the flue route (Figure 7), especially in the replacement market, the flue position may be individually selected.

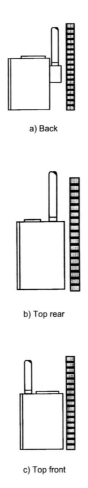

a) Back

b) Top rear

c) Top front

Figure 7 Position of flue connections

Also available are two room sealed versions:

- Se-duct/U-duct

- balanced flue.

Many modern ducted air heaters combine a circulator to serve a cylinder providing hot water for the property. Information on circulators can be found under *Heating Boilers/Water Heaters*.

BUILDING HEAT LOSS AND COMFORT CONDITIONS

The heating requirements for each room in the dwelling should be calculated in the same way as for wet central heating systems. From these calculations the correct rating of ducted air heater and duct sizes can be selected.

DUCTS

Air is blown across the heat exchanger by the fan, picking up heat, and continues into the plenum box. The plenum box, is in effect, the manifold from which warm air is distributed to the ducts (Figure 8). The burners supplying the heat exchanger of the unit are isolated from the effects of air movement created by the fan. Should the heat exchanger fail, i.e. perforate or split, the burner flames may/will be disturbed when the fan operates creating possible incomplete combustion and sooting within the combustion chamber. Similarly products of combustion could enter the duct system prior to operation of the fan and then be distributed around the dwelling when the fan operates. A similar situation can occur where the warm air unit is not sealed to the plenum box.

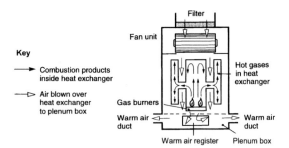

Figure 8 Typical ducted air heater

The warm air ducts terminate either through grilles (referred to as diffusers) at ceiling or floor level, or through wall-mounted grilles (referred to as registers) which are normally adjustable with an on/off lever.

By fitting restrictor plates behind the diffusers or registers, the heat output at each grille can be balanced in much the same way a lockshield valve is used on a radiator.

Ducts should be run and installed in a secure manner, joints should be made with pop-rivets or self-tapping screws and all joints sealed with duct tape. The fitting of ducts requires the installer to be competent in sheet metal work. Ducts should be insulated with a minimum of 50 mm glass fibre or equivalent to prevent against heat losses exceeding 2°C per 3 m length.

Insulation within 2 m of the heater should be thermally stable up to 120°C.

TYPES OF DUCT SYSTEM

Stub Duct System

This type of system is generally found on older installations, the ducted air heater being sited in a central position within the dwelling, with short ducts (stubs) communicating with adjoining rooms.

Usually, in this type of installation the appliance is fitted in a compartment; this compartment so positioned that each of its four sides communicates with a different room, e.g. lounge, dining room, hall and kitchen. The ducted air heater may also have a short vertical duct that communicates with one or two upper bedrooms.

In local authority housing, a minimum standard was set when the Parker Morris Report 1961 laid down minimum acceptable temperatures for domestic heating.

During the house construction boom in the 1960s and 1970s many local authorities interpreted these as standards and not as minimum temperatures. This resulted in thousands of homes being built with partial central heating, e.g. the stub duct warm air system, usually with no heating to the bedroom area.

The manufacturers of ducted air heaters nowadays are able to provide information to help overcome the heating deficiencies of these earlier types of system. This may mean replacing the ducted air heater with one more suited to the task, and where necessary extending the ducting system to include bedrooms.

Fully Ducted System

This type of system has a network of heating ducts installed to supply warm air to each room of the property. The ducts are suitably sized to provide the correct level of heat to each room.

Fully ducted air systems can be subdivided:

- extended plenum system
- radial warm air system
- stepped duct system.

The various systems may use a combination of square, rectangular and circular ducts. Figures 9, 10, and 11 show typical layouts for the three systems.

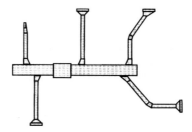

Figure 9 Extended plenum – maximum length of any branch duct 6 m

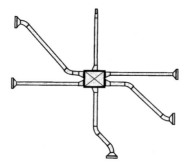

Figure 10 Radial warm air system – maximum length of any duct 6 m

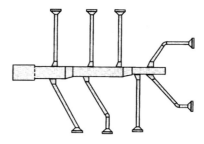

Figure 11 Stepped duct system – maximum length of any branch duct 6 m

RETURN AIR PATHS

So far this chapter has dealt with air being blown into a room, through the appropriate size duct to a register or diffuser.

BS 5854 requires a full and unobstructed return air path to the appliance from all heated rooms and spaces within a property, with the exception of kitchens, bathrooms, toilets and shower rooms.

Open Flued Appliances

The return air arrangement shall be such that suction from the circulating fan does not interfere with the operation of the flue.

Where it is necessary to install a purpose-provided air vent (non-closable type), to compensate for any effect that the fan may have on the flue, the following guidance is given.

Where the property is two storeys or less, either a relief air grille (supplementary ventilation from outside the property) or transfer grille (ventilation via one internal heated room) to the collection area is acceptable.

Where the property is more than two storeys, only a relief air grille direct from outside to the collection area is acceptable.

For maisonettes or flats in multi-storey buildings, a relief air grille direct from outside, to the collection area, should be installed wherever practically possible. Where this is not reasonably practical, then a transfer grille will have to be installed from a heated room into the collection area.

Transfer grilles must not be installed more than 450 mm above floor level (to prevent the passage of smoke or flames in the event of a fire).

When examining this type of system, all return air paths from heated rooms to the heater collection area must be checked. This is to determine that they are correctly sized and installed.

British Standards require that the return air grille from the collection area into the compartment must be ducted through the compartment (open flued heaters only), and be connected onto the appliance. This duct must be sealed to prevent intermixing of return air and combustion air.

Where another open flued appliance is installed in the collection area, then that appliance, as well as the ducted air heater, must be tested to confirm adequate removal of products of combustion.

Tests must be carried out with both appliances in operation.

Any other fans, such as ducted cooker hoods, vented tumble dryers and room extract fans, anywhere in the property, will require consideration in accordance with current flue testing procedures.

Positive Return Air

Since 1970, all open flued ducted air heaters installed should have a ducted return air path (sealed through the compartment where the appliance is fitted), connected to the appliance.

It is important that the arrangement of the ducted return air path is installed to prevent the following conditions:

- the action of the fan having an adverse effect on open flue performance, resulting in flue reversal

- products of combustion being circulated around the property, if the flue fails to work effectively

- flue failure induced by an excessively long uninsulated flue or a partially blocked flue, resulting in spilled products of combustion.

Positive ducted return air has the benefit of providing not only a tolerance of *flue safety* as outlined above, but also:

- assists control of ventilation rates throughout the property

- has the ability to add to the system efficiency by introducing a heat recovery unit into the property, and sending recovered heat back through the return air duct to the appliance.

Correctly calculating and siting the return air provision is important to provide an effective warm air system. The return air grilles, located within most rooms of the property, need to be positioned in relationship to the direction of air flow and the positioning of the warm air delivery grilles, to avoid complaints of underheating caused by short cycling between the grilles.

POSITIVE RETURN AIR SHOULD NOT BE TAKEN FROM A ROOM CONTAINING OPEN FLUED APPLIANCES.

RETURN AIR GRILLES MUST BE OF THE NON-CLOSABLE TYPE.

Types of Return Air Connection to the Ducted Air Heater

There are two methods commonly used for connecting the return air duct to the appliance (Figure 12). One is a rigid metal duct and the other is a non-combustible flexible connection.

There is a danger that the appliance may be insecure if the flexible connection is used, and additional fixings may be necessary during installation to secure it adequately.

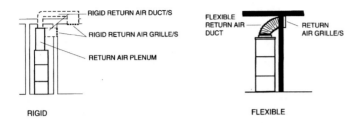

Figure 12 Return air path connections

Whichever duct connection method is used, the return air duct must be sealed. This is to ensure that the combustion air provision within the compartment cannot intermix with that from the return duct and that the heater fan cannot interfere with the operation of the flue.

For adequate property ventilation it is possible to introduce a supply of outside air. This can be achieved by running a duct that communicates with outside atmosphere either via a ventilated roof space or grille on an outside wall into the return air duct system. Under these circumstances the duct that carries the incoming air should be adjustable to give a minimum flow rate of 2.2 m^3/h for every kW of appliance input, this adjustment being achieved by incorporating a lockable damper into the air inlet duct.

COMBUSTION VENTILATION

The same rules apply as for any other gas appliance. Open flued ducted air heaters with a heat input in excess of 7 kW net will require a vent sized at 5 cm^2 per kW of appliance maximum rated input net. Further guidance can be found in the ConstructionSkills publication *Gas Safety* or BS 5440-2.

COMPARTMENT VENTILATION

Open Flued

The same rules apply as for any other gas appliance.

Purpose-provided air vents must be provided at high and low level to provide both combustion air and circulation air to dissipate the heat from the compartment.

Room Sealed

Purpose-provided air vents at high and low level are required, again in keeping with the aforementioned Regulations and Standards, to dissipate the heat from the compartment.

Any gas appliance installed in a compartment must have an appropriate warning notice.

DUCTED AIR HEATER CONTROLS

FLAME SUPERVISION DEVICES

Thermo-electric

In its simplest form, the thermo-electric device (thermocouple) is a loop of two dissimilar metals joined together at one end, with the other ends connected to an electro-magnet. When the joint or junction is heated, a small voltage is produced (see figure 13).

The voltage produced is dependent on the temperature and the metals used. Generally thermocouples used as flame supervision devices utilise a chrome-nickel alloy and copper. The output voltage produced for these metals is between 15 to 30 mv. When the joint is heated by a pilot flame, the voltage energises the magnet thus holding the armature to it in a spring-operated gas valve and allowing gas to flow to the main burner.

Should the pilot be extinguished, the thermocouple would cool down and stop producing a voltage, thus allowing the spring to close the valve.

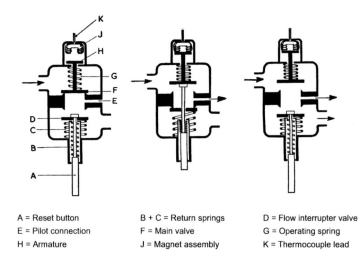

A = Reset button   B + C = Return springs   D = Flow interrupter valve
E = Pilot connection   F = Main valve   G = Operating spring
H = Armature   J = Magnet assembly   K = Thermocouple lead

Figure 13 Thermo-electric flame supervision device

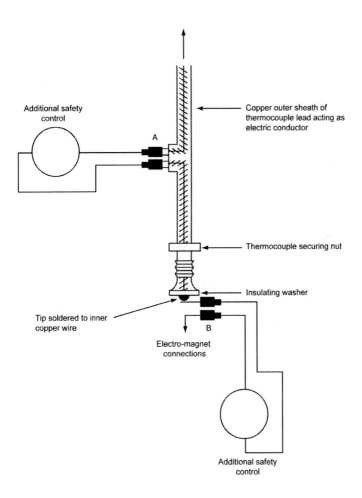

**Figure 14 Interruptible thermocouple
(connections made at either position A or B)**

Faults

- **Pilot flame**
  Partially blocked or incorrect position of pilot flame, which should play on the tip or the top 12 mm of the thermocouple. Under-aerated pilot flames are easily disturbed by draughts and have a low flame temperature.

- **Thermocouple**
  Contacts must be clean and tight (tight means about a quarter of a turn beyond hand tight). Short circuits can be caused by distortion of insulating washer by overtightening of the union nut. The tip should be clean and undamaged. The tip should be kept below red heat if the thermocouple is to have a reasonable life.

- **Electromagnet**
  Failure rarely occurs of the armature and magnet, which are housed within a sealed unit. Unit exchange is required when this happens.

Testing thermo-electric flame supervision devices (FSD)

There are numerous procedures for determining whether these devices have shut down correctly. The factors which determine which procedure to adopt are:

- the availability of pressure testing nipples on the equipment involved

- the accessibility of the burner or pilot burner being controlled (i.e. open or room-sealed flues)

- whether the FSD shut-off can be proved by including the procedure within a tightness test of the complete gas installation using the test nipple on the gas meter.

Where the manufacturer's instructions do not specify a procedure then the following may be used:

1. Ensure that all primary and secondary controls are set so that the burner will not be turned off during this procedure.

2. Light the appliance, and allow the burner to reach its normal working temperature.

3. Turn off the appliance shut-off device (appliance isolation valve) and simultaneously start a stopwatch.

4. Halt the stopwatch when the valve is heard to close.

5. Immediately check that the valve in the FSD has shut off completely, using the most appropriate method as indicated below:

    - **Preferred option**
      Where the appliance gas control system has a test nipple upstream of the FSD device (normally defined as test point P1), test for tightness and let-by between the appliance isolation valve and the FSD. If the FSD device has not shut off completely then a drop in pressure will occur.

- **Option 2**
  Where it is possible to complete a tightness test at the meter, or other suitable position upstream of the appliance isolation valve, then by turning on the appliance isolation valve the integrity of the FSD is also included. (Procedures must be adopted to ensure that any escape indicated by the gauge is not elsewhere on the gas installation pipework other than the FSD.)

- **Option 3**
  Where neither of the above options are available and the appliance is open flued, or the pilot and main burner are readily accessible, then where possible connect a gauge to the burner test nipple and turn on the appliance isolation valve. If any apparent increase in pressure is observed immediately turn off the isolation valve as this indicates that the FSD has not shut off completely. If no apparent increase in pressure is observed (or no gauge has been connected) immediately check with a lighted taper that gas has been interrupted to the main and pilot burner.

6. Check that the time recorded by the stopwatch conforms with the current requirements for gas appliances of heat inputs below 70 kW.

Oxygen Depletion System (ODS) – Atmosphere Sensing Device (ASD)

The European Gas Directive, 1 January 1996, states that when undergoing type testing to obtain the CE mark, appliances connected to a flue for the dispersal of combustion products must be so constructed that in abnormal draught conditions there is no release of combustion products in a dangerous quantity into the room concerned.

Domestic gas appliance design allows for excess air under normal operating conditions to be entrained into the appliance combustion chamber and hence to the atmosphere via the flue. When there is a spillage of combustion products into the room where the appliance is installed, complete combustion will occur for a period even though the oxygen level is decreasing and the carbon dioxide level is rising. However, as the oxygen level falls further, incomplete combustion occurs and carbon monoxide (CO) starts to be produced. The appliance design is such that the rate of CO production is initially low as the oxygen level falls and it is at this point that the oxygen depletion system (ODS) within the appliance intervenes.

A typical ODS (S.I.T. Gas Controls Limited) uses a controlled flame to heat a thermocouple, being part of a thermo-electric flame supervision device. As the oxygen level decreases in the atmosphere, so this controlled flame 'lifts' in search of oxygen, thus reducing the heat applied to the tip of the thermocouple until at a pre-determined point the electric current is reduced sufficiently to shut off the gas supply to the appliance (figures 15, 16 and 17).

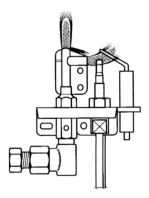

Figure 15 Adequate oxygen supply

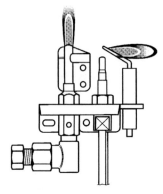

Figure 16 As the oxygen level falls, the sensing flame lifts away from the thermocouple tip

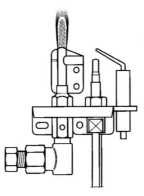

Figure 17 Just prior to shutdown – the sensing flame has completely extinguished

The ODS has an intervention level of 200 ppm (0.02%) of CO concentration in the room in which the appliance is installed.

The installation and annual servicing of all appliances must be conducted by competent operatives and those checks and tests to prevent incomplete combustion occurring as specified in Regulation 26(9) of the current Gas Safety (Installation and Use) Regulations must be complied with.

The ODS device must be checked according to the manufacturer's instructions whenever work has been carried out on the appliances in addition to any annual safety checks. The main points of these checks are to ensure:

- no part of the ODS is damaged

- the ODS is securely mounted in its recommended location

- the flame picture is not distorted and is burning correctly at the main burner cross-ignition port, the sensor port and its inter-connecting ribbon burner

- the aeration port adjacent to the injector is free from any obstruction.

If a customer reports that the ODS keeps 'going out' there is a high probability that it is working correctly and doing exactly what it is supposed to do by making the appliance safe in the event of progressive oxygen starvation due to abnormal flue or ventilation conditions.

Note: The S.I.T. Oxypilot ODS has no serviceable components and if required a complete unit exchange is necessary (figure 18).

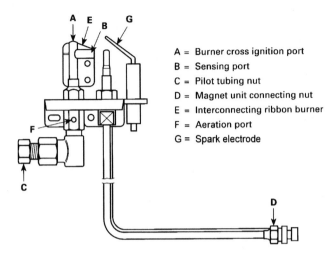

A = Burner cross ignition port
B = Sensing port
C = Pilot tubing nut
D = Magnet unit connecting nut
E = Interconnecting ribbon burner
F = Aeration port
G = Spark electrode

Figure 18 S.I.T. Oxypilot ODS

Flue Safety Devices

These devices are used to detect adverse flue conditions (spillage) at the draught diverter of an open flued appliance. They are known as TTB's (a Dutch acronym of the words 'Themische Tervgslag Beveiliging') but are often referred to as down draught thermostats, thermoswitches or smoke thermostats.

These devices (heat sensors) are located just inside the draught diverter, and are linked:

- in series with the thermocouple of a thermoelectric flame supervision device, or
- to shut off the main burner solenoid valve, or
- to the electronic circuit board.

The sensors are pre-set and calibrated to avoid nuisance shutdowns while still maintaining safe tolerances. They require manual intervention to re-establish the gas supply to the main burner.

Flame Rectification

This method of flame supervision superseded the more basic flame conductance system, which was prone to simulated d.c. flame signals. Condensation or a build up of carbon, due to flame chilling, can bridge the probe and burner. With a d.c. signal where electrons travel around the circuit in only one direction, a control unit can not distinguish the presence of a flame, from the bridging of the gap between the probe and the burner. However an a.c. signal can, due to the two directional flow to and from the control unit.

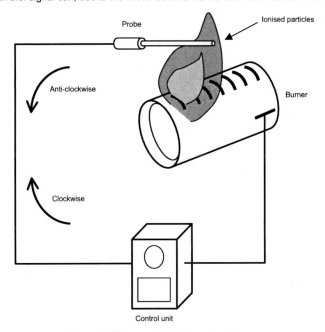

Figure 19 **Flame rectification circuit**

If we imagine the electrons from the control unit's signal, travelling in a clockwise direction, when the signal reaches the probe, the electrons are able to travel to the burner due to the ionised particles in the gas flame. If there was no flame present, the electrons are not supplied with sufficient pressure to jump the gap e.g. voltage/spark. The probe passing the electrons to the burner is very much like a shotgun firing pellets at a barn door (the burner has a much greater area). Therefore all of the electrons will travel the gap and be registered at the end of the clockwise journey back to the control unit.

When travelling back in the anti-clockwise direction, the electrons now try to pass the gap from the burner to the probe, this is now like shooting a cannon at a pencil, only some of the electrons are able to 'hit' the probe and travel back to the control unit. We now find a rectified signal recognised by the control unit as the presence of a flame.

The flame rectification system can distinguish various signals, for example:

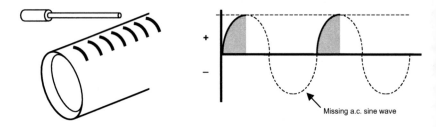

Figure 20 Open circuit

Figure 20, shows the signal read by the control unit where no flame or bridge is present. The electrons reach the end of the probe but have nowhere to go.

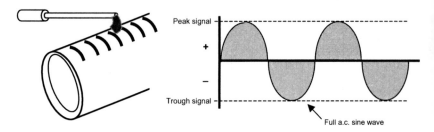

Figure 21 Closed circuit

Figure 21, shows the signal read by the control unit when the gap between the probe and burner, is bridged by conductive matter (condensation or carbon). The electrons can travel freely in both directions.

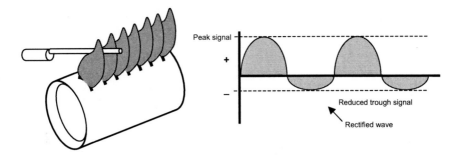

Figure 22 Rectified signal

Figure 22, shows the signal read by the control unit when the gap between the probe and burner, is bridged only by the flame, the shot gun and barn door effect now takes place, rectifying the signal.

A typical, burner head to probe ratio of 8:1 will allow a rectified signal to be produced. This ratio is easily surpassed for most atmospheric bar type burners.

Fused Power Supply

This is the main point of electrical isolation for the ducted air heater and its associated controls. It may be a 3-pin plug or a fused spur type connection, located adjacent to the heater.

The electrical supply requires a 3 amp fuse, this point must always be 'switched off' and isolated, i.e. fuse out/plug out before any work is carried out on the heater.

Timing Controls (Clocks and Timers)

Timing controls give the facility to preset a heating system with 'on' and 'off' periods at predetermined times, repeating on a 24 hour cycle.

Timing devices can be supplied by the appliance manufacturers fitted and wired integrally within the appliance, thereby reducing the need for some of the external wiring. Alternatively they can be fitted remote from the appliance and wired accordingly.

Unlike timers for wet central heating, timers used on ducted air heaters normally switch low voltage i.e. 24 V. It should however be noted that manufacturers are in the process of producing some new models which switch 240 V.

Room Thermostat

A room thermostat offers automatic adjustment of room temperature by directly controlling the ducted air heater. It consists of a mechanical sensing device, an electrical switch and a temperature selector dial. The sensing devices are either bimetallic strips/coils, vapour-filled bellows, or double diaphragms. Room thermostats for ducted air heaters normally operate at 24 V.

The room thermostat should be situated in a position that represents as far as possible, the average heating requirements of the house. Temperature variations from room to room should be catered for in the initial system design and balancing. Room thermostats should be located away from misleading factors such as heat sources, direct sunlight, or localised draughts. They should be fitted 1.5 m above floor level.

Limit Thermostat

This is a safety thermostat fitted integrally within the ducted air heater which shuts down the burner (via a solenoid valve) if an overheat condition occurs on the heater, such as fan failure, blocked/dirty air filter etc.

The limit thermostat is normally set at 82–92°C (180–200°F).

Fan Thermostat

The fan thermostat mechanism can be adjusted but in general the following settings are used. 'Fan off' normally set to approximately 38°C (100°F) and 'differential' which is set to 17°C (30°F) or 'fan on' setting which will bring the fan on at 55°C (130°F).

The fan thermostat controls the fan in the following way:

The burner is ignited and when the temperature in the heat exchanger reaches 55°C (130°F) ('fan off' plus 'differential') the fan starts. When the room thermostat is satisfied, the burner shuts off and the fan continues until the heat exchanger temperature cools to 38°C (100°F), then the fan shuts off. This disperses the residual heat from the heat exchanger, discharges useful heat to the rooms, prevents overheating of the appliance and the resultant possible tripping of the overheat thermostat.

Some fan thermostats have a summer/winter override switch which enables the fan to be used on its own without heat.

Summer/Winter Switch

This simple, manually operated, switch is fitted in the main voltage supply to the fan. In the 'winter' position the switch is open and the fan thermostat controls the fan. If the user wishes to run the fan when the heating unit is not operating, the switch is changed to 'summer' position. The fan will then circulate air through the same rooms that would normally be heated.

Fan

The fan is powered by a 240 V supply and circulates the air over the heat exchanger and around the duct system.

Transformer

This is fitted within the ducted air heater and steps down the voltage from 240 V to 24 V for such components as gas valves and, on older appliances, the limit thermostat.

Wiring

Conventional ducted air heaters are very simple to wire up as they utilise basic heater controls with the addition of a 24 V circuit being the only other consideration. Figure 23 illustrates a typical wiring diagram of a conventional ducted air heater.

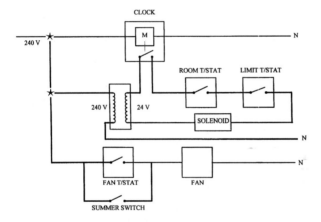

Figure 23 Ducted air heater – functional flow wiring diagram

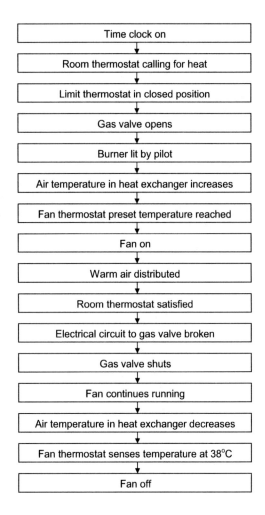

Figure 24 Sequence of operation of a basic ducted air heater

MODAIRFLOW

Johnson and Starley introduced this new system of controlling ducted air heaters in the 1970s.

The Modairflow control automatically adjusts heat output, and modulates the flow of warm air, in balance with heat requirements. Stable room temperatures result from the continuous gentle circulation of just the right amount of warm air at comfortable flow temperatures.

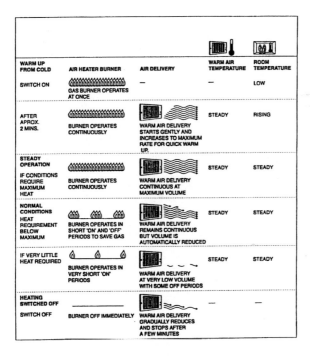

Figure 25 Operation of the Modairflow system

Control of the Modairflow System

Special controls, built into the ducted air heater, continuously adjust its operation to maintain the chosen comfort level. This system delivers a continuous supply of warm air at just the right flow to keep room temperature steady. It works smoothly, quietly and avoids perceptible fluctuations of airflow and temperature.

Time Control

Modulating airflow systems only require a simple time control which can be set to supply the required 'heat on/off' times.

Thermista-stat

The Thermista-stat continuously checks the actual temperature in the room, compares it with the dial setting, and instructs the ducted air heater to deliver the necessary volume of warm air. The dial is set to the comfort number required; if the temperature setting is reduced the system will react by burning less fuel but will continue to run. An electronic control unit within the ducted air heater analyses the heat requirements sensed by the Thermista-stat and programmes the heater and burner accordingly for a period of operation between 2 and 3 minutes. At the end of this period the heater is re-programmed with adjustments made according to the heat requirements.

The high sensitivity of the Thermista-stat (combined with the sophisticated response of the modulating airflow system) enables room temperature to be controlled to within ¼°C, compared with about 2°C achievable with traditional on/off systems.

Wiring

The wiring of a modulating airflow system is very similar to a conventional system utilising simple controls and a mixture of 240 V and 24 V. An additional consideration to take into account is the Thermista-stat, which operates on DC current and therefore it is important to ensure that the correct polarity is adhered to. This is indicated both on the unit and the Thermista-stat itself. Figure 26 illustrates a typical wiring diagram for a modulating airflow system (Modairflow control unit).

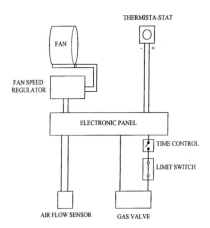

Figure 26 Modairflow control unit

INSTALLATION OF DUCTED AIR HEATERS

Advice on materials, design and installation of ducted air heaters is given in BS 5864: Installation and maintenance of gas-fired ducted warm air heaters of rated input not exceeding 70 kW net.

Before installing any gas appliance, the manufacturer's instructions should be consulted. They give specific directions, including location requirements and minimum dimensions with regard to flue terminal positions.

For some appliances you may find the measurements given differ slightly from those stated in the British Standards. In this case the manufacturer has designed the appliance to operate satisfactorily with regard to specific installation instructions which should be adhered to.

The installation must also conform to the current Gas Safety Regulations and, if the installation contains an electrical supply, to the Institute of Electrical Engineers Regulations (IEE).

The choice of a particular type of ducted air heater should match the requirements of the customer, taking into account acceptable positions for installation and comparative installation and maintenance costs.

LOCATION AND INSTALLATION GUIDELINES

Ducted air heaters can be installed in a number of different positions. Often the layout of the building dictates the site.

There are general conditions that prevail and must be maintained in the siting of the ducted air heater.

1. Products of combustion must be effectively discharged to outside air.

2. A sufficient supply of air should be available for combustion and to give adequate ventilation of the heater compartment. Conventionally flued heaters should be so sited that under no circumstances can the discharge of products of combustion pass into a bedroom or bathroom as a result of downdraught. When the installation of a ducted air heater in a bedroom is considered, a room sealed appliance only should be used.

3. Wherever an appliance is located, any local byelaws, fire regulations or insurance company's requirements shall be complied with.

4. Delivery duct runs between heater and registers or grilles should be as short and direct as possible.

5. An adequate, unobstructed return air path to the heater should be available, except from kitchens and WCs.

6. The sound level of operation should not cause annoyance. Ducted air heaters should not generally be sited in living rooms or bedrooms.

7. Adequate space should be allowed for access to burner, filter, pulleys, motor, flueways, controls and other parts requiring regular attention.

8. Adequate clearance must be allowed between the heater or its flue and any combustible material.

9. Where the heater is installed on a suspended floor the mounting employed should be such as to prevent the floor acting as a sound board, e.g. resilient mounting or padded concrete slab.

FLUES AND VENTILATION

An adequate supply of air is essential for a gas appliance to burn safely and efficiently. For open flued appliances an air supply is also required to ensure adequate ventilation in the room or space in which the appliance is located.

The air requirement for a given appliance is dependent upon its rated heat input, whether it is open flued or room sealed, and its location in a room or a compartment. Room sealed ducted air heaters are permissible in any room or internal space provided that the flue terminal can be located as recommended by the manufacturer, or BS 5440-1.

The following gives guidance on acceptable locations for a ducted air heater with regard to flueing and ventilation as recommended by BS 5864.

More detailed information on flueing and ventilation of gas appliances can be found in the ConstructionSkills publication *Gas Safety*.

Bathrooms and Shower Rooms

Under no circumstances shall any open flued ducted air heater be installed in a room containing a bath or shower.

Only room sealed appliances may be fitted in these locations, provided that the flue terminal can be located as recommended in BS 5440.

Where a room sealed ducted air heater, with an electrical supply, is located in a room containing a fitted bath or shower, the mains electrical switch should be so sited that it cannot be touched by a person using the bath or shower, in accordance with the Institute of Electrical Engineers Regulations.

Living Rooms, Kitchens, Utility Rooms, Halls and Passageways

When selecting a living room as a location for a ducted air heater, careful consideration should be given to the effect on the living area of such an installation, e.g. noise of the appliance, general aesthetics of the installation and servicing requirements.

The type of appliance that can be fitted is unrestricted providing there is sufficient ventilation and an adequate flue system is available.

Compartments and Cupboards

A compartment is defined as an enclosed space within a building, either constructed or modified specifically to accommodate the ducted air heater and its ancillary equipment.

Where a compartment is used to accommodate a ducted air heater the compartment should conform to the following requirements:

- be of a rigid structure in accordance with any manufacturer's instructions regarding the internal surfaces

- combustible internal surfaces should be at least 75 mm (3 in) from any part of the heater, or be suitably protected with a non-combustible material

- must have access to allow inspection, servicing and removal of the appliance and ancillary equipment

- the compartment must incorporate air vents for ventilation and, where necessary, for combustion as recommended in BS 5440-2

- no air vent must communicate with a bathroom, shower room, bedroom or bedsitting room if an open flued appliance is fitted in the compartment

- the return air grille(s) shall be directly connected to the return air inlet on an open flued ducted air heater

- the door shall be self-closing to a compartment housing an open flued ducted air heater if the door communicates with a bedroom or bedsitting room

- the door must have a warning notice attached, stating that the door must be kept closed except for when resetting the appliance controls.

Airing Cupboards

A warning label should advise the customer of the potential dangers of storing, drying and airing clothes in the compartment housing a ducted air heater.

Any airing space should be separated from the ducted air heater compartment by a non-combustible material which may have perforation holes, provided they are not greater than 13 mm (½ in) across any axis. This prevents any clothing falling onto the heater.

The flue pipe should not pass through the airing space unless protected by a guard, such as wire mesh, that prevents stored linen being placed within 25 mm (1 in) of the flue pipe.

Under Stairs Cupboards

This location should only be considered as a last resort in a building that is more than two storeys high. Wherever possible only room sealed appliances should be installed. Air vents must be direct to the outside air.

The cupboard should be treated as a compartment and, in addition, in a building that is more than two storeys high all the surfaces should be lined with non-combustible material unless inherently fire resistant, e.g. plastered ceiling.

Slot-fit Installations

Only specifically designed appliances for slot fit application should be installed in this way. The manufacturer's instructions should be complied with.

The area directly above the appliance should be enclosed to prevent the obstruction of the draught diverter, combustion air inlets and return air inlets.

Roof Space Installations

Where no other alternative exists and where local regulations permit, an open flued or room sealed appliance may be fitted.

When this location is selected the roof space should have:

- flooring of sufficient strength and area to support the appliance and facilitate servicing
- enough vertical clearance when siting the cold water cistern to ensure the availability of the static head required by a water heating appliance
- a suitable means of access to the heater, e.g. foldaway loft ladder, and sufficient fixed lighting
- a guard fitted to prevent any items stored within the roof space coming into contact with the appliance
- a guard-rail around the access hatch
- gas isolation control outside of the loft area.

COMMISSIONING CHECKLIST FOR DUCTED AIR HEATERS

It should be noted that the checklist shown may be used in the absence of the appliance manufacturer's instructions.

On completion of the installation, the gas supply should be tested for tightness and purged, and a flue flow test applied to any open flued system. The commissioning procedure below may then be followed.

1. Ensure installation meets all relevant regulations.
2. Check ventilation is adequate, where appropriate.
3. Ensure the return air path is clear and adequate.
4. Ensure the filter, fan and fan compartments are free from obstructions.
5. All registers and grilles are open and conform to design specifications.
6. Ensure all controls are in the off position.
7. Check ignition device and light pilot.
8. Check pilot flame length and position, and operation of flame supervision device.
9. Turn on electricity supply, time control and room thermostat.
10. Ensure main burner has ignited.
11. Ensure that the fan stat starts the fan about 1 to 2 minutes after ignition.
12. Check the burner pressure corresponds to data badge.
13. Check the gas rate, if necessary, and the flame picture.
14. Carry out spillage test. See spillage test procedure (Figure 27 and Figure 28).
15. Ensure all controls operate satisfactorily:
    - overheat/limit thermostat, by closing all registers
    - fan switch
    - room thermostat.

16. When these checks are complete the system will need to be balanced as follows:
    - ensure that the ducted air heater operates for about 20 minutes so that the temperature can stabilise
    - check that temperature rise across the heater is between 45°C and 55°C
    - set the fan speed accordingly (decrease fan speed to increase temperature)
    - measure air velocities at each register or diffuser and set the dampers to provide the required velocity.

Finally the customer should be instructed in the use of the appliance including lighting, operating and the cleaning of the filter.

MAINTENANCE CHECKLIST FOR DUCTED AIR HEATERS

It should be noted that the checklist shown may be used in the absence of the appliance manufacturer's instructions.

PRE-SERVICE CHECKS

1. Check with the customer to ascertain any problems or faults with the installation.
2. Check the general condition of the appliance and that the installation conforms to appropriate Standards.
3. Check that there is adequate ventilation and that the flue is both routed and terminated correctly.
4. Check there are no signs of spillage on the appliance or adjacent walls.
5. Check the operation of all controls, the flame supervision device and ignition system.
6. Check flame picture.
7. Advise the customer of any problems.

FULL SERVICE

1. Isolate gas and electricity.
2. Dismantle as necessary and clean dust and deposits from within the casing.
3. Check for signs of damage to electrical connections, cables or components, clean and rectify as necessary.
4. Remove and clean main burner and injectors.
5. Remove and clean pilot burner and injectors.
6. Remove and clean the fan and air filter.
7. Inspect the heat exchanger and flueways. Carry out a flue flow test on open flued systems.
8. Ease and grease gas taps, as necessary.
9. Check all appliance and ductwork seals and joints.
10. Check all disturbed gas connections.

11. Test the ignition device by lighting the heater.
12. Check pilot flame and test flame supervision device.
13. Ensure that the working pressure, gas rate and flame picture are correct.
14. Check operation of controls as fitted, including:
    - overheat/limit thermostat, by closing all registers
    - fan switch
    - room thermostat
15. Check operation of all electrical controls.
16. Carry out spillage test. See spillage test procedure (Figure 27 and Figure 28).
17. Leave the appliance in working order and advise the customer of any further work required.

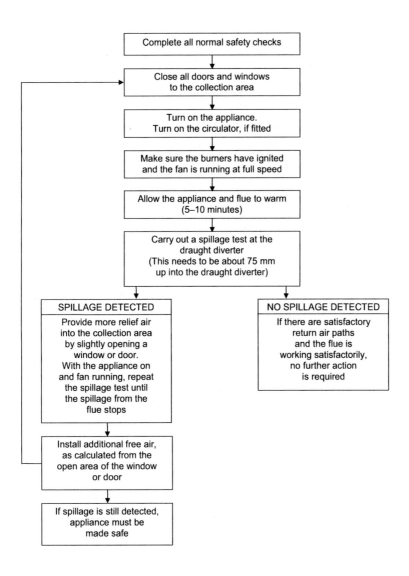

**Figure 27 Spillage testing on ducted air heaters –
draught diverter is accessible**

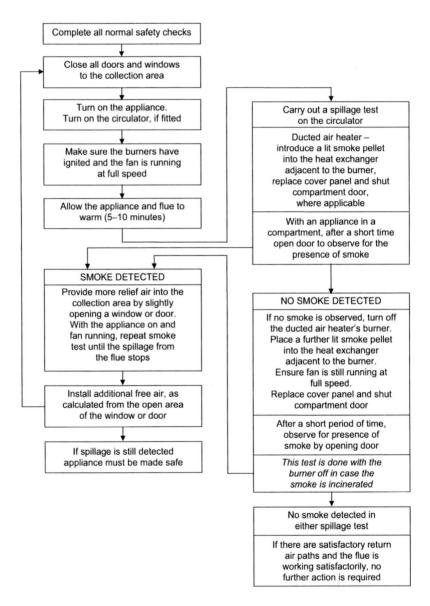

**Figure 28 Spillage testing on ducted air heaters –
draught diverter is not accessible**

FAULT-FINDING CHECKLIST FOR DUCTED AIR HEATERS

Symptom	Possible cause	Remedy
1. Main burner lights but fan fails to run after a short period	a) Loose electrical connection on fan b) Fan control settings incorrect c) Faulty fan assembly d) Faulty fan control e) Burner setting pressure not correct	a) Check connections for tightness b) Check settings c) Replace, taking care not to damage impeller d) Replace e) Adjust pressure, as necessary
2. Main burner operating intermittently with fan running	a) Gas rate or burner setting pressure high b) Temperature rise excessive c) Air filter or return air path restricted d) Excessive number of outlets closed	a) Check gas rate and burner setting pressure b) Adjust fan speed or gas rate accordingly c) Check filter is clean and return air path is clear d) Open additional outlets
3. Main burner operating with intermittent fan operating	a) Gas rate or burner setting pressure low b) Fan control settings incorrect	a) Check gas rate and burner setting pressure b) Check settings
4. Fan runs for excessive period or operates intermittently after main burner shuts down	a) Fan control settings incorrect	a) Check settings
5. Noisy operation	a) Gas pressure high b) Noisy fan motor c) Fan speed setting too high	a) Check burner setting pressure b) Replace fan motor c) Adjust fan speed

Fires and Wall Heaters

CONTENTS

	Page
INTRODUCTION	1
The Gas Appliances Directive	1
Gas Appliances (Safety) Regulations 1995	1
European Standards	2
Regulations and Standards Affecting Installation and Maintenance	3
Health and Safety at Work etc. Act 1974 (HSW Act)	3
Management of Health and Safety at Work Regulations 1999 (MHSWR)	3
Reporting of Injuries, Diseases and Dangerous Occurrences Regulations 1995 (RIDDOR 95)	3
Building Regulations 2000 and Building (Scotland) Regulations 2004	3
British Standards	4
Manufacturer's Instructions	5
SPACE HEATERS	6
Heat Transfer	6
Radiant Heat	6
Convected Heat	7
Conducted Heat	7
Types of Space Heater	7
Open Flued Gas Fires	7
Room Sealed Gas Fires	8
Flueless Room Heaters	8
Balanced Flue Convectors	8
GAS FIRES	10
Radiant Fires	10
Radiant Convector Fires	10
Glass-fronted Fires	12
Free Standing Heating Stove	13
Electrically Simulated Log/Coal-effect Fires	14
Balanced Flue Radiant Convector Fires	14
Room Sealed Fan-flued and Closed Flue Gas Fires	14
Decorative Fuel Effect Gas Fires	15
Inset Live Fuel Effect Gas Fires	16
Fanned Draught Flue Systems	16
Condensing Appliances	17
Component Parts of Gas Fires	17
Outer Case	17
Dress-guard	17
Firebox	17
Radiants	17

(continued overleaf)

	Page
Imitation Logs and Coals	19
Firebrick	19
Heat Exchanger (for Radiant Convector)	19
Burners	19
Simplex Burner	19
Duplex Burner	20
Consumer Control	21
Controls	21
Ignition	21
Flame Supervision Device and Oxygen Depletion System (ODS)	21
Oxygen Depletion System (ODS) – Atmosphere Sensing Device (ASD)	21
Closure Plate	25
Levelling Adjusters	25

FLUELESS SPACE HEATERS (CONVECTORS)   26

BALANCED FLUE CONVECTORS   27
    Natural Draught, Natural Convection   27
    Fanned Draught, Fanned Convection   28

INSTALLATION OF GAS FIRES AND CONVECTOR HEATERS   29
    Materials and Components   29
    Preliminary Examination   29
    Second-hand Appliances   29
    Location – Permanent Dwellings   29
    Ventilation   30
       Convectors   30
       Open Flued Appliances   30
       Multi-appliance Installations   31
    Flues   33
       Masonry Chimneys   33
       Precast Flue Block Chimneys   34
       Sheet Metal Flues   34
       Catchment Spaces   34
    Fire Precautions   36

FLUE SIZING FOR DECORATIVE FUEL EFFECT
(NATURAL DRAUGHT) FLUE SYSTEMS   46
    Flue Termination for Decorative Fuel Effect Fires   48
       Terminals and Chimney Pots   48
       Location   48
       Bird Guards   48

FLUE PROVING   51
    Pre-installation   51
    Flue Testing   51

PREPARATION OF THE APPLIANCE   52
    Gas Supply   52

(continued overleaf)

	Page
INSTALLING AND COMMISSIONING GAS FIRES CHECKLIST	55
Method of Testing for Spillage	57
INSTALLATION OF CONVECTOR HEATERS	58
Balanced Flue Appliances	58
Location	58
Flueless Space Heaters	58
Gas Supply	58
Appliance Fixing	58
Ventilation	59
MAINTENANCE OF SPACE HEATERS	60
Open Flued Heaters	60
Preliminary Inspection	60
Service	60
Room Sealed Heaters	61
Preliminary Inspection	61
Service	62

INTRODUCTION

This introduction contains an appraisal of some of the most important Regulations that determine both the design and manufacture of certain domestic appliances.

To reinforce this, there is reference to the Regulations and British Standards that domestic gas appliance installations must comply with. These Regulations and Standards also impose certain constraints on the subsequent maintenance of appliances and the pipework installation.

THE GAS APPLIANCES DIRECTIVE

Since 1985 the European Union has been pursuing a 'common approach' to tackle the problems associated with trade barriers between member states. For example, EU countries have different laws at present relating to product safety. Removing these barriers is at the heart of the Single European Market introduced in 1992.

The Gas Appliances Directive sets out legal requirements that in future will apply across the European Union. Member countries are required to amend their existing legislation, or to introduce new legislation that conforms with the requirements of the directive. The United Kingdom has implemented the Gas Appliances (Safety) Regulations to conform with the directive.

GAS APPLIANCES (SAFETY) REGULATIONS 1995

Until 1992, the safety (to consumers) of gas appliances sold in the United Kingdom has been covered by the Consumer Protection Act and specifically by the Gas Cooking (Safety) Regulations, and the Heating Appliances (Fireguards) Regulations. The Gas Appliances (Safety) Regulations 1995 introduced specific requirements.

There was a transitional period until 1996 in which gas appliances offered for sale in the United Kingdom were allowed to meet the old requirements.

The main provision of the new Regulations are:

a) **appliances must be safe**

b) **appliances must be tested**

c) **appliances must be quality guaranteed.**

This means that during the manufacturing process the manufacturer must operate a quality scheme of some type, such as BS 5750, to ensure that all appliances conform to the tested design. This scheme will be monitored by the 'Notified Bodies'.

d) appliances must carry the CE mark

All appliances that conform to provisions **a), b) and c)** will carry a CE mark issued by the 'Notified Bodies' (see Figure 1).

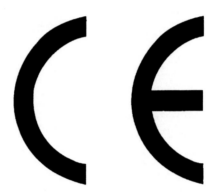

Figure 1 CE mark

The Regulations include detailed procedures for product conformity attestation by third party notified bodies, appointed by the Secretary of State.

All new gas appliances must have information included, covering safe installation, operation and maintenance.

EUROPEAN STANDARDS

European Standards are currently being compiled. For some appliances, where no European Standard is planned, the National Standards (in this country, British Standards) may be recognised. This, for example, will apply to the British type of gas fire.

REGULATIONS AND STANDARDS AFFECTING INSTALLATION AND MAINTENANCE

Health and Safety at Work etc. Act 1974 (HSW Act)

This Act applies to everyone concerned with work activities, ranging from employers, self-employed, and employees, to designers, suppliers and importers of materials for use at work, and people in control of premises. The duties apply both to individual people, and to corporations, companies, partnerships, local authorities etc. Employers have a duty to ensure, so far as is reasonably practicable, the health, safety and welfare at work of all employees, and not to expose people who are not their employees to risks to their health and safety.

Management of Health and Safety at Work Regulations 1999 (MHSWR)

These Regulations impose a duty on employers and self-employed persons to make suitable and sufficient assessment of risks to the health and safety of employees, and non-employees affected by their work. It also requires effective planning and review of protective measures, health surveillance, emergency procedures, information and training.

Reporting of Injuries, Diseases and Dangerous Occurrences Regulations 1995 (RIDDOR 95)

These Regulations require employers to report specified occupational injuries, diseases and dangerous occurrences (events) to the HSE. Certain gas incidents are reportable by suppliers of gas through fixed pipe distribution systems and/or LPG suppliers, and gas installers are required to report certain dangerous gas appliances to the HSE.

Building Regulations 2000 and Building (Scotland) Regulations 2004

These Regulations address the various aspects of building design and construction which include energy conservation and health and safety.

The Secretary of State has approved a number of documents under the Building Regulations 2000 as practical (non-mandatory) guidance to meeting the requirements under the Regulations.

Similar 'deemed to satisfy' guidance is provided in technical handbooks of the Building (Scotland) Regulations 2004.

The documents that particularly relate to gas work in domestic premises are:

- **Building Regulations 2000 (England and Wales)**

    Part A – Structure

    Part B – Fire Safety

    Part F – Ventilation

    Part G3 – Hot Water Storage

    Part J – Combustion Appliances and Fuel Storage Systems

    Part L – Conservation of Fuel and Power

    Part M – Access To and Use of Buildings

    Part P – Electrical Safety

- **Building (Scotland) Regulations 2004**

    Section 1 – Structure

    Section 2 – Fire

    Section 3 – Environment

    Section 4 – Safety

    Section 6 – Energy

BRITISH STANDARDS

British Standards' specifications are an invaluable guide to the installation of gas appliances. If followed, these standards will satisfy the requirements of current Regulations.

The following is a selection of some of the important British Standards Specifications relating to Domestic Gas Appliances, which give guidance on the minimum standard that appliance installations should comply with, to satisfy current Regulations:

BS 5546:	2000	– Specification for installation of gas hot water supplies for domestic purposes
BS 5588:	(Domestic) Part 1 1990	– Fire precautions in the design, construction and use of buildings
BS 5864:	2004	– Installation and maintenance of gas fired ducted warm air heaters of rated input not exceeding 70 kW net (2nd and 3rd family gases)
BS 5871:	Part 1 2005	– Gas fire, convector heaters and fire/back boilers (2nd and 3rd family gases)

BS 5871: Part 2 2005 – Inset fuel effect gas fires of a heat input not exceeding 15 kW (2nd and 3rd family gases)

BS 5871: Part 3 2005 – Decorative fuel effect gas appliances of a heat input not exceeding 20 kW (2nd and 3rd family gases)

BS 6700: 1997 – Design, installation, testing and maintenance of water supplies for domestic purposes

BS 8423: 2002 – Fire-guards for fires and heating appliances for domestic use - Specification

BS 6172: 2004 – Installation and maintenance of domestic gas cooking appliances (2nd and 3rd family gases) – Specification

BS 6798: 2000 – Specification for installation of gas fired boilers of rated input not exceeding 70 kW net

BS 6891: 2005 – Installation of low-pressure gas pipework of up to 35 mm (R 1¼) in domestic premises (2nd family gases)

BS 5440: Part 1 2000 – Flues

BS 5440: Part 2 2000 – Air supply

BS 5482: Part 1 2005 – Code of practice for domestic butane and propane gas burning installations. Installations in permanent buildings, residential park homes and commercial premises up to 28 mm

BS 7624: 2004 – Installation and maintenance of domestic direct gas fired tumble dryers up to 6 kW heat input (2nd and 3rd family gases) - Specification

MANUFACTURER'S INSTRUCTIONS

Manufacturer's instructions are important for the installation, commissioning, maintenance and use of any gas appliance. These instructions must be read and followed.

After installation of the appliance or subsequent maintenance of it, the instructions must be returned to the consumer so that they may store them for future reference. This includes both user and installation/servicing instructions. (This is a requirement of the Gas Safety Regulations.)

SPACE HEATERS

Individual space heaters offer the householder an important alternative to full central heating, providing a controllable, efficient appliance that can deliver heat when needed. Most situations can be catered for by the large range of appliances available. These can be installed to heat part of a property, such as a lounge, or they can provide heating to a number of rooms and other spaces.

The main benefits are their flexibility, offering:

- quick warming
- low running costs
- low installation costs
- independent control.

Many households with central heating also install a space heater in the lounge to:

- create a focal point in the room
- supplement the central heating
- provide a standby in the event of the heating system becoming faulty
- provide a quick response to chilly evenings during late summer.

HEAT TRANSFER

If space heating is being used as an alternative to central heating, heat loss calculations will be necessary to determine the heat requirements needed to ensure comfort conditions for the property.

There are three different methods of heat transfer. It is important that the difference between radiant, convected and conducted heat is understood.

Radiant Heat

Radiant heat is the heat that is transmitted through air, or through a gas. The sun, for example, heats the earth by radiant heat. Radiant heat travels in straight lines and only heats the objects that are directly in its path. With a gas fire the two main benefits of this type of heat transfer are:

- an immediate feeling of warmth
- the psychological feeling of warmth, when looking at the heated radiants, or coals.

Convected Heat

This is the term given to the transfer of heat by currents of air. Hot air rises, cooler air replaces it and creates movement of air. Heat is distributed throughout the room or space by this method. It can in certain circumstances cause the sensation of draughts.

Conducted Heat

This is the term given to the transfer of heat by touch. Conduction is the major factor in a building losing heat. Hot will always travel to cold and in the case of poorly insulated houses, the conducted heat loss through walls, floors, ceilings and windows may be considerable.

TYPES OF SPACE HEATER

This section considers the different types of space heater concentrating on their general design, installation and servicing requirements.

The term 'space heater' encompasses all kinds of heating appliances including:

- open flued gas fires
- room sealed gas fires
- flueless space heaters (convectors)
- balanced flue convectors.

Open Flued Gas Fires

Radiant (typical rated heat input 5–6 kW)

Radiant convector fires (typical rated heat input 4–8 kW)

Decorative fuel effect fires (typical rated heat input 7–18 kW)

Inset gas fires (typical rated heat input 7–15 kW)

Room Sealed Gas Fires

Room sealed radiant/convector gas fires
(typical rated heat input 4–8 kW)

Wall heaters (typical rated heat input 2–15 kW)

Flueless Room Heaters

Hall heaters (typical rated heat input 1.5–2 kW)

Balanced Flue Convectors

Typical rated heat input 2–15 kW

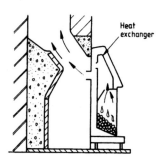

Flue size: A minimum of 125 mm across the axis of the flue is normally required.

Location: Normally in front of the closure plate which is fitted to the fireplace opening.

Ventilation: Purpose provided ventilation is not normally required up to 7 kW input.

Note: For this type of appliance the radiating surface can be in the form of either a radiant(s) or an imitation fuel, the latter giving a live fuel effect.

a) Gas fire (to BS 7977-1)

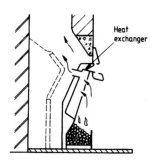

Flue size: A minimum of 125 mm across the axis of the flue is normally required.

Location: It is either fully or partially inset into the builder's opening or fireplace recess. (For a recess, the chairbrick may have to be removed, depending upon the appliance design.)

Ventilation: Purpose provided ventilation is not normally required up to 7 kW input.

b) Inset live fuel effect gas fire (ILFE) (to BS 7977-1)

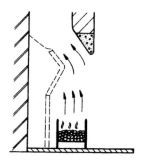

Flue size: A minimum of 175 mm across the axis of the flue is normally required.

Location: Within the builder's opening or fireplace recess or flue box, or under an associated independent canopy. (For a recess, the chairbrick may have to be removed, depending upon the appliance design.)

Ventilation: In the absence of manufacturer's instructions, purpose provided ventilation of at least 100 cm^2 is normally required up to 20 kW net heat input.

c) Decorative gas fire (DGF) (to BS EN 509)

Figure 2 Types of gas fire

GAS FIRES

RADIANT FIRES

This is the simplest form of open flued fire. Approximately 95% of its heat is distributed to the room by radiant heat. A typical efficiency for this type of appliance is 50%. (See Figure 3.)

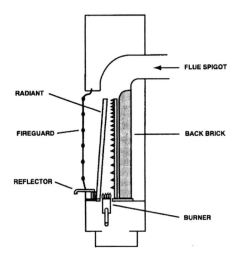

Figure 3 Radiant gas fire

RADIANT CONVECTOR FIRES

The introduction of this type of fire represents one of the most important developments in open flued gas appliances in the UK. During the 1950s the radiant convector gas fire gave the gas industry a tremendous boost, with an appliance that was both efficient and economical to run.

By introducing a heat exchanger above the radiant area, and directing products of combustion through this before entering the flue, the efficiency of the gas fire is increased to around 65%. Heated air leaves the heat exchanger through case louvres and is replaced by cooler air, flowing into the fire from hearth level which, in turn, is heated to provide a continuous cycle. In addition, objects in the room are heated by radiant heat.

The main advantages of this type of appliance, over the radiant gas fire, are not just the increase in efficiency, but also a better distribution of heat and a more comfortable living environment. (See Figure 4.)

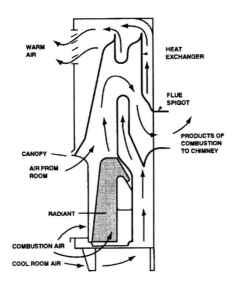

Figure 4 Radiant convector

GLASS-FRONTED FIRES

Many gas fires are manufactured with heat resistant glass fronts (Figure 5). The combustion process is contained behind a glass panel. They may be balanced flue or open flue.

It is important for installers to be able to distinguish between the different flue systems.

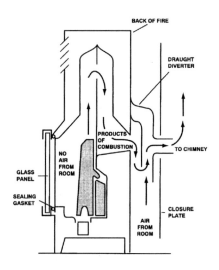

Figure 5 Glass-fronted fires

FREE STANDING HEATING STOVE

A variation of the glass-fronted stove, this appliance (Figure 6) is available with either a top outlet flue where the appliance is set into a recess or a rear outlet when fitted in conjunction with a closure plate and catchment space on a free standing hearth in front of the fireplace opening.

When fitted into the recess in conjunction with a lined flue (Figure 7) it is important to ensure that the short length of flue attached to the appliance protrudes at least 150 mm above the register plate to ensure any debris that may fall down the flue does not block it.

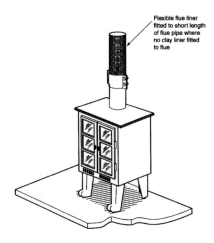

Figure 6 Freestanding stove

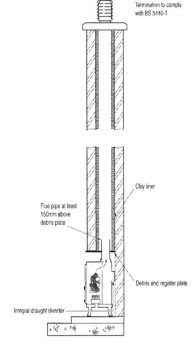

Figure 7 Installation of freestanding stove

ELECTRICALLY SIMULATED LOG/COAL-EFFECT FIRES

Many fires are manufactured with a semi-translucent moulding to imitate logs or coals. This is illuminated by electric bulbs to enhance the visual 'warmth' of the appliance.

BALANCED FLUE RADIANT CONVECTOR FIRES

The room sealed gas fire with a heat-resistant glass front was introduced to overcome the problem of properties having no chimney/flue. The flue and siting of the flue terminal should be installed and checked in accordance with BS 5440-1 and to manufacturer's instructions.

The heat is distributed throughout the room in an identical method to its open flued equivalent. The radiants or imitation coal/logs are located behind the glass panel and still emit radiant heat. A heat exchanger above the radiants produces convected heat that circulates through the case louvres into the room.

The front glass panel is normally removable allowing access to the radiants and burner. As with any other balanced flue appliance the seals on this panel need to be sound. (See Figure 8.)

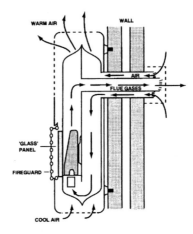

Figure 8 Balanced flue radiant convector

ROOM SEALED FAN-FLUED AND CLOSED FLUE GAS FIRES

There have been many interesting developments in recent years with the introduction of room sealed fan-assisted flue radiant convector gas fires and condensing fires.

DECORATIVE FUEL EFFECT GAS FIRES

An altogether different appliance from those previously explained in this chapter. Appliances of this type are classified as only being suitable for installing to a class 1 flue. This means that they are to be treated, from a flue size point view, as if they are a solid fuel fire. The installer is also normally required to provide ventilation.

These appliances have become very popular as they mimic, visually, the effect of solid fuel or log-burning fires.

The main disadvantage of these appliances is their inefficiency. As they have little or no control on the heat being lost up the chimney the efficiency is reduced. Typical efficiency for basket or simple inset is approximately 20%. (See Figure 9 and Figure 10.)

Figure 9 Decorative fuel effect

Figure 10 Inset decorative fuel effect gas fire

INSET LIVE FUEL EFFECT GAS FIRES

This gas fire is inset into the builder's opening of a chimney flue. It is designed with a heat exchanger above the imitation logs/coal to improve the efficiency. (See Figure 11.)

Efficiency for inset live fuel effect gas fires is approximately 40–45%.

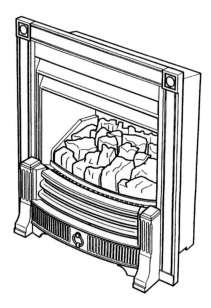

Figure 11 Inset live fuel effect gas fire

FANNED DRAUGHT FLUE SYSTEMS

Only those appliances where the appliance manufacturer's instructions state that they are suitable shall be installed to a fanned draught flue system. The flue installation must be in accordance with the appliance and flue manufacturer's instructions.

There must be an interlock (fan flow device) to shut down the appliance in the event of failure of the flue draught.

Condensing Appliances

These normally incorporate a fanned draught flue system. Special jointing techniques may apply, and any instructions should be followed. Unless otherwise specified, there should be a continuous drop in height of the flue from the appliance to the termination of 1:40, and the terminal should project a minimum of 75 mm away from the point of exit to avoid condensation staining the wall.

COMPONENT PARTS OF GAS FIRES

Outer Case

The case of a gas fire may have either a strictly functional design, with enamel or paint finish, or an aesthetically pleasing look with alternative finishes in wood veneer, copper or brass.

On radiant/convector fires there is a grill built into the top of the case assembly to allow the convected heat to circulate.

Dress-guard

Dress-guards are a requirement of current Regulations and Standards to prevent the accidental contact, by touch, with the incandescent material (radiants). Radiant and radiant convector gas fires manufactured in the UK are fitted with a dress-guard as standard equipment.

Where young children, elderly or infirm people have access to a gas fire, the consumer should be advised that an additional fireguard, conforming to BS 8423, be fitted to the front of the chimneybreast around the appliance (see also page 37).

Firebox

This is the part of the fire that houses the radiants. The products of combustion are directed through the firebox to the flue.

Radiants

Radiants are manufactured for individual fires and generally are not interchangeable between appliances. Radiants are made of fireclay and are fragile. (Figure 12)

Care should be taken to follow the manufacturer's instructions when fitting the radiants to the fire after installation or servicing.

Wrongly positioned radiants may lead to:

- an obstruction of secondary combustion air
- flame impingement
- sooting
- possible spillage of combustion products.

Radiants are normally supported by a plate above the burner or by hooks moulded into the rear of the radiant, which locate on the firebox.

The flames from the burner heat the internal surface of the radiants until it glows. The products of combustion pass through the opening at the top of the radiants into the canopy of the firebox, through the heat exchanger and out to the flue. The typical operational temperature of a radiant is approximately 900°C – this must be considered if removing radiants from a fire shortly after use.

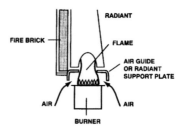

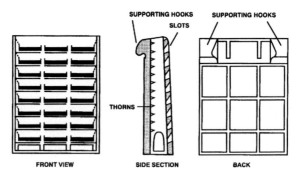

Figure 12 Radiants

Imitation Logs and Coals

Many modern gas fires have replaced the radiant with a set of imitation logs or coals. These are made of various materials such as fireclay, volcanic lava or ceramic. The burner fires through the bed of logs/coals to produce the visually pleasing effect of a real log or coal fire.

The manufacturer's instructions are normally very explicit about the arrangements of these logs/coals and care must be taken to ensure correct positioning, otherwise severe sooting may occur.

By design, these appliances produce yellow flames and therefore small amounts of soot deposits. Coal/log-effect gas fires need to be operated at full gas rate on initial lighting and for a period of time before turning off. This is to help to burn off any soot deposits. Specific instruction, regarding any particular appliance in this category, will be found in the manufacturer's instructions.

Firebrick

Most gas fires are manufactured with refractory bricks behind the radiants. These firebricks minimise heat loss through the rear of the fire and are secured to or supported on the firebox.

Heat Exchanger (for Radiant Convector)

Heat exchangers for these types of appliance may be made from a variety of materials. Normally, coated pressed steel or in some cases cast iron is used. They absorb some of the heat still contained in the combustion products leaving the radiants before the combustion products pass into the flue, and therefore raise the heater's efficiency.

Burners

Pre-aerated burners are used for these types of appliance. They may be of either simplex or duplex design.

Simplex Burner

This type of burner is fed by one injector (Figure 13).

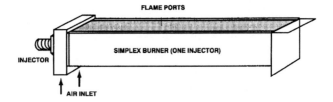

Figure 13 Simplex burner

Duplex Burner

This type of burner is fed through two or more injectors. The burner comprises two burners within one unit, thereby enabling the outer radiants to be controlled independently of the centre radiants (Figure 14 and Figure 15).

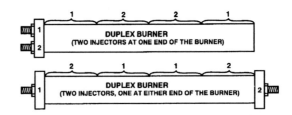

Figure 14 Duplex burner

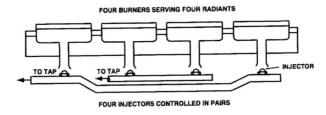

Figure 15 Duplex burner

Consumer Control

The majority of control taps will be of the plug and taper type, with holes through the plug to give the desired heat input to a simplex burner or duplex burner.

Controls

The construction, operation and fault diagnosis of controls are detailed in the ConstructionSkills publication *Gas Safety*.

The main controls used on gas space heaters are as outlined below.

Ignition

High-voltage or Spark Ignition

There have been many ignition methods used for gas fires over the years. Most fires use high-voltage spark ignition. This commonly is achieved by piezo-electric crystal.

Any spark ignition device needs a sound electrical connection from the generator to the electrode and will require the electrode gap to be adjusted correctly (3 to 5 mm).

Pilot Light Ignition

Many gas fires are manufactured with a pilot light ignition system. These may be used just for automatic ignition, or be incorporated within the design for flame supervision purposes.

Flame Supervision Device and Oxygen Depletion System (ODS)

Many gas fires are fitted with a flame supervision device and/or oxygen depletion system. Such safety devices should fail to safety in accordance with British Standards.

Oxygen Depletion System (ODS) – Atmosphere Sensing Device (ASD)

The European Gas Directive, 1 January 1996, states that when undergoing type testing to obtain the CE mark, appliances connected to a flue for the dispersal of combustion products must be so constructed that in abnormal draught conditions there is no release of combustion products in a dangerous quantity into the room concerned.

Domestic gas appliance design allows for excess air under normal operating conditions to be entrained into the appliance combustion chamber and hence to the atmosphere via the flue. When there is a spillage of combustion products into the room where the appliance is installed, complete combustion will occur for a period even though the oxygen level is decreasing and the carbon dioxide level is rising. However, as the oxygen level falls further, incomplete combustion occurs and carbon monoxide (CO) starts to be produced. The appliance design is such that the rate of CO production is initially low as the oxygen level falls and it is at this point that the oxygen depletion system (ODS) within the appliance intervenes.

A typical ODS (S.I.T. Gas Controls Limited) uses a controlled flame to heat a thermocouple, being part of a thermo-electric flame supervision device. As the oxygen level decreases in the atmosphere, so this controlled flame 'lifts' in search of oxygen, thus reducing the heat applied to the tip of the thermocouple until at a pre-determined point the electric current is reduced sufficiently to shut off the gas supply to the appliance (figures 16, 17 and 18).

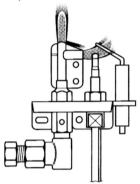

Figure 16 Adequate oxygen supply

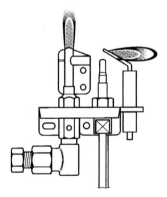

Figure 17 As the oxygen level falls, the sensing flame lifts away from the thermocouple tip

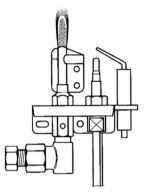

Figure 18 Just prior to shutdown – the sensing flame has completely extinguished

The ODS has an intervention level of 200 ppm (0.02%) of CO concentration in the room in which the appliance is installed.

The installation and annual servicing of all appliances must be conducted by competent operatives and those checks and tests to prevent incomplete combustion occurring as specified in Regulation 26(9) of the current Gas Safety (Installation and Use) Regulations must be complied with.

The ODS device must be checked according to the manufacturer's instructions whenever work has been carried out on the appliances in addition to any annual safety checks. The main points of these checks are to ensure:

- no part of the ODS is damaged
- the ODS is securely mounted in its recommended location
- the flame picture is not distorted and is burning correctly at the main burner cross-ignition port, the sensor port and its inter-connecting ribbon burner
- the aeration port adjacent to the injector is free from any obstruction.

If a customer reports that the ODS keeps 'going out' there is a high probability that it is working correctly and doing exactly what it is supposed to do by making the appliance safe in the event of progressive oxygen starvation due to abnormal flue or ventilation conditions.

Note: The S.I.T. Oxypilot ODS has no serviceable components and if required a complete unit exchange is necessary (figure 19).

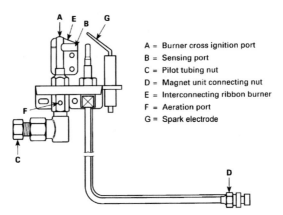

A = Burner cross ignition port
B = Sensing port
C = Pilot tubing nut
D = Magnet unit connecting nut
E = Interconnecting ribbon burner
F = Aeration port
G = Spark electrode

Figure 19 S.I.T. Oxypilot ODS

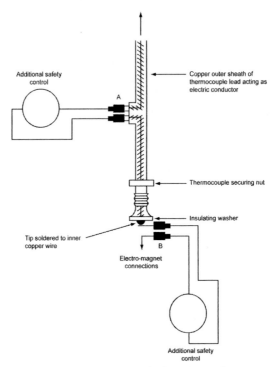

**Figure 20 Interruptible Thermocouple
(connections made at either position A or B)**

Closure Plate

The closure plate plays an essential part in the safety and the efficiency of radiant and radiant convector gas fires. They must be of the correct dimensions, which are determined by the manufacturer.

The closure plate performs a number of very important functions which include:

- providing the correct air change rate to the room
- ensuring that the products of combustion are satisfactorily vented to the flue system
- providing access to the base of the flue for servicing purposes, and
- providing positive location for the flue spigot of the fire to enter the flue system.

The installation of such a plate, sealed on all four sides, is essential and the apertures cut into it must be of the precise measurements detailed by the manufacturer. Closure plates are provided by the manufacturer in the case of new fires. However, second-hand appliances supplied without a closure plate will require the installer to construct a suitable plate, from a material such as sheet aluminium, in accordance with the measurements in the manufacturer's instructions.

Levelling Adjusters

Care should be taken to ensure the appliance is properly levelled. Follow the levelling arrangements detailed in the manufacturer's instructions.

FLUELESS SPACE HEATERS (CONVECTORS)

These are of simple construction consisting of a burner located within a sheet steel box forming a combustion chamber. The warm air and combustion products are discharged through louvres (within an outer case of steel and wood) into the room or space in which the appliance is installed.

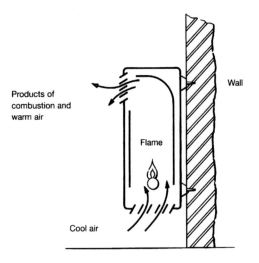

Figure 21 Fixed flueless convector

The controls to these appliances are normally a plug and tapered control tap and an ignition device. As it is classified as a continuous burning flueless appliance, the heat input is normally limited to 1.5 kW or 2.0 kW. The heat inputs allowed into rooms or spaces from these appliances are limited according to the room or space volume.

BALANCED FLUE CONVECTORS

NATURAL DRAUGHT, NATURAL CONVECTION

These are balanced flue appliances consisting of a room sealed heat exchanger of usually coated pressed steel or cast iron construction, connected to a balanced flue set terminating outside.

Non-electric models are normally available incorporating a thermoelectric flame supervision device within a multifunctional control, piezo ignition and possibly a regulator and a thermostat.

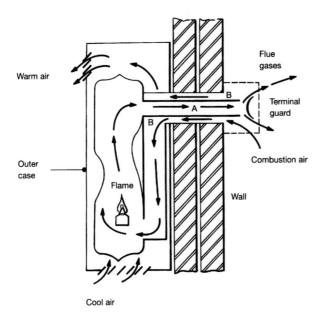

Figure 22 Balanced flue convector – natural draught, natural convection

FANNED DRAUGHT, FANNED CONVECTION

These appliances have fan-assisted room sealed heat exchangers and flue systems. Both intake of air and discharge of products are through a balanced flue terminal. The air intake and flue ducts are a lot smaller than natural draught appliances (approximately 70 mm compared to 140 mm diameter). By utilising a fan to force cool air from the room over the appliance heat exchanger, the appliance heat output is greatly increased.

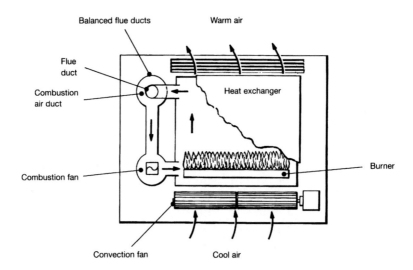

Figure 23 Balanced flue convector, fanned draught, fanned convection, separate fan motors

The controls used on fan-assisted heaters vary considerably from one model to another. Protection is normally required against overheating if the convector outlet louvres become blocked, and an interlock between the combustion fan and the gas supply to the burner protects against fan failure. Full sequence control systems including pre-purge of the heat exchanger may also be available.

INSTALLATION OF GAS FIRES AND CONVECTOR HEATERS

MATERIALS AND COMPONENTS

All materials and components used in the installation shall comply with the requirements of the applicable British Standards or, where no British Standard exists, materials and equipment should be of a suitable quality and workmanship to fulfil their intended purpose. Materials containing asbestos shall not be used.

PRELIMINARY EXAMINATION

The appliance data badge shall be examined to ensure that the gas supplied and the operating pressure are suitable,

    i.e.   natural gas   G20

           propane   G31

           butane   G30

SECOND-HAND APPLIANCES

When a second-hand appliance is installed, or when an appliance is moved from one location to another, the physical condition of the appliance should be checked to ensure that the appliance is free from deterioration, distortion or displacement of any component that may adversely affect the safe condition of the appliance.

LOCATION – PERMANENT DWELLINGS

Open flued appliances shall not be installed in a private garage or in a room or internal space containing a bath or shower.

Open flued gas fires combined with a back-boiler or back-circulator shall not be installed in a bedroom or bed-sitting room.

Flueless convector appliances shall not be installed in a private garage, bedroom, bed-sitting room, or in a room containing a bath or shower.

Where flueless appliances are installed in rooms, the rated net heat input shall not exceed 45 W/m^3 of heated space in the room, or 90 W/m^3 of heated space when installed in other locations such as halls or passages.

Room sealed appliances may be installed in any room or space providing it complies with BS 5440-1 and 2 and the appliance manufacturer's instructions.

An appliance for use with 3rd family gases (i.e. propane or butane) shall not be installed in a room or internal space below ground level, e.g. in a basement or cellar. This does not prevent the installation where the room or space are basements with respect to one side of the building, but which are open to ground level on the opposite side.

VENTILATION

The ventilation requirements for all types of space heaters shall be in accordance with BS 5440-2 for Natural and LP Appliances Installed in Permanent Dwellings, except for Decorative Fuel Effect Fires, where the ventilation requirements are specified in BS 5871-3.

Convectors

a) A room sealed appliance does not require an air vent in the room or internal space in which it is installed.

b) A flueless convector has its ventilation requirements determined from the maximum permissible net input allowed according to the volume of the room or space in which it is installed.

To convert gross input to net input use the following:
Net input = gross input ÷ 1.11 for natural gas; 1.09 for propane and 1.08 for butane.

The ventilation requirements are:

- fixed flueless space heater installed in a room:
    - maximum net input = 45 W/m^3 of heated space
    - purpose provided ventilation = 100 cm^2 for net inputs up to 2.7 kW, plus 55 cm^2 for each kW input exceeding 2.7 kW net

- fixed flueless space heater installed in an internal space:
    - maximum net input = 90 W/m^3 of heated space
    - purpose provided ventilation = 100 cm^2 for net inputs up to 5.4 kW, plus 27.5 cm^2 for each kW input exceeding 5.4 kW net.

In any location, in addition to any purpose provided ventilation, there must be an opening window direct to outside. Acceptable alternatives include any adjustable louvre, hinged panel or other means of ventilation that opens directly to outside air.

Open Flued Appliances

a) Radiant convector fires with a rated net or gross input not exceeding 7 kW do not normally need an air vent in the room or space in which they are installed, unless the appropriate flue spillage test indicates that additional ventilation is required. An air vent shall not communicate directly with a builder's opening or fireplace recess.

b) Inset live fuel effect fires with a rated net or gross input not exceeding 7 kW do not normally need an air vent in the room or space in which they are installed unless otherwise stated in the manufacturer's instructions, or if the appropriate flue spillage test indicates that additional ventilation is required.

Where a fire has been tested to the appropriate standard and such tests have shown that the clearance flue flow to avoid spillage exceeds 70 m^3/h, purpose provided ventilation will be necessary. The manufacturer's instructions will provide details in such circumstances.

For appliances with a rated net heat input exceeding 7 kW, purpose provided ventilation of at least 5 cm^2/kW of heat input above 7 kW shall be provided up to 15 kW, unless otherwise specified in the manufacturer's instructions. An air vent shall not communicate directly with a builder's opening or fireplace recess.

c) Decorative fuel effect fires with a rated net input not exceeding 20 kW will require a purpose provided ventilation opening of not less than 100 cm^2 unless otherwise specified in the manufacturer's instructions.

An air vent shall not communicate directly with a builders opening or fireplace recess. Where Radon gas has been identified as a problem, ventilation shall not be taken from below floor level using a floor vent. Floor vents communicating with a ventilated underfloor void are permitted and may be used in other cases provided they are outside of the hearth area.

Multi-appliance Installations

a) Open flued radiant convector and inset live fuel effect gas fires may require purpose provided ventilation if other gas appliances are also installed in the same room or space. The resultant air vent requirements should be the LARGEST of the following:

- the total flueless space heating appliance requirements

- the total open flued space heating requirements (gas fires, central heating boilers, warm air units, convectors or the total of a combined appliance such as fire/backboilers or ducted air heaters/water heaters)

- the largest individual requirement of any other type of appliance.

Where these fires are installed in a through room, (where an interconnecting wall has been removed between the two rooms, and the resultant room contains two chimneys, each terminating at the same height and preferably close together), and either fire is installed to each of these chimneys, then a purpose provided air vent is NOT required if the rated heat input of each appliance does not exceed 7 kW net or gross.

b) Decorative fuel effect gas fires

   i) Where a room or internal space contains one of these gas fires and one or more additional gas appliances, the purpose provided air vent requirement shall be the total requirement for the decorative fuel effect fire PLUS, where appropriate, the GREATEST of the following:

   - the total flueless space heating appliance requirement, or

   - the total open flued space heating appliance requirement (as defined under *Multi-appliance Installations* (a) above) WITH NO ALLOWANCE FOR ADVENTITIOUS AIR, or

   - the greatest individual requirement of any other type of appliance, and if open flued non-space heating appliance WITH NO ALLOWANCE FOR ADVENTITIOUS AIR.

   Where two decorative fuel effect gas fires are installed in a room or internal space, 35 cm^2 is to be added to the total air vent free area requirement determined above.

   ii) Through-room installations involving one decorative fuel effect (DFE) gas fire and other gas appliances connected to a similar flue

   Where the room contains only one DFE fire, the purpose provided air vent requirement shall be the total DFE fire requirement (see *Open Flued Appliances* (c) above), PLUS where appropriate, the greatest of the following:

   - the total flueless space heating appliance requirement, or

   - the total other open flued space heating requirement ALLOWING FOR ADVENTITIOUS AIR (i.e. 5 cm^2 per kW in excess of 7 kW net), or

   - the greatest individual requirement of any other type of gas appliance, and if open flued non-space heating appliance ALLOWING FOR ADVENTITIOUS AIR (i.e. 5 cm^2 per kW in excess of 7 kW net).

   iii) Through-room installations involving two decorative fuel effect gas fires and other gas appliances connected

   Where the room contains two similar decorative fuel effect gas fires, the purpose provided air vent requirement shall be the total DFE requirement (see *Open Flued Appliances* (c) above), PLUS where appropriate the greatest of the following:

   - the total flueless space heating appliance requirement, or

   - the total open flued space heating appliance requirement WITH NO ALLOWANCE FOR ADVENTITIOUS AIR (i.e. 5 cm^2 per max. kW input net), or

   - the greatest individual requirement of any other type of appliance. For open flued non-space heating appliance the requirement shall be with NO ALLOWANCE FOR ADVENTITIOUS AIR (i.e. 5 cm^2 per max. kW input net).

c) Multi-appliance installations involving other fuels

The installation of any gas fire in the same room, space, or through room as a solid fuel open fire, or an appliance with an open fan-assisted flue is NOT RECOMMENDED. (Ref. Gas Safety (Installation & Use) Regulations 1998 Regulations and Guidance 26 (1) in HSE Document L56.)

Where existing installations contain these appliances, or oil-fired appliances, they shall be treated as if they were gas appliances of similar type and rated input. Where only the output rate is available the input should be calculated using the following equation:

$$\text{Input} = \text{output} \times \frac{10}{6}$$

For an existing solid fuel open fire or a small closed stove of unknown heat input the purpose provided air vent requirement should be taken as 100 cm^2.

d) Effect of extract fans

Extract fans that may effect the ventilation requirement for any open flued appliance fitted in the same room or in an adjacent room to where the appliance is installed, must be checked.

FLUES

With the exception of flueless heaters, all gas fires will discharge their products of combustion to external air, and shall be flued in accordance with BS 5440-1. Before installing or working on any fire, the correct operation of the flue shall be verified in accordance with BS 5440-1 and with the HSE Approved Code of Practice and Guidance to the Gas Safety (Installation and Use) Regulations 1998. Specifically, Regulation 27 and Appendix 1 should be referred to.

Masonry Chimneys

Pre-1966 solid fuel type chimneys were normally 9 in x 9 in brick built with an inner skin skimmed with mortar. During previous use the mortar skin can deteriorate, and falling mortar and soot can block or restrict the chimney. The resulting lack of the skin of mortar can cause leaking brick joints, resulting in poor flue performance and wall staining.

Gas fires (except direct connection types) may be connected to this type of flue without lining it, providing the flue is in a satisfactory condition and its length does not exceed 10 m external wall or 12 m internal wall. Gas fires and convectors with a direct flue connection should not be connected to this type of flue unless a flexible flue liner is fitted or the manufacturer's instructions permit it.

Post-1996 solid fuel type chimneys were lined with square or round clay pipes, or with concrete liners. It is important that these liners are sealed at the bottom to the surrounding brickwork to produce a satisfactory flue pull.

Any type of open flued gas fire may be connected to this type of satisfactory solid-fuel chimney.

Any chimney which has previously been used for an appliance burning a fuel other than gas shall be swept before the gas fire is installed.

Any damper or restrictor plate in the chimney shall be removed, or where it is not reasonably practical to remove a sliding damper, it shall be permanently fixed in the fully open position.

Before installing any fire the base of the flue shall be cleared of debris.

Precast Flue Block Chimneys

Various types of pre-cast flue blocks are available. The fire manufacturer's instructions should be consulted regarding the suitability of a fire for use with pre-cast flue block systems, and any special instruction appertaining to the particular appliance. Where a cooler device is required it should be specified or supplied by the fire manufacturer.

Sheet Metal Flues

These shall comply with BS EN 1856 or BS 715. A flue box shall only be used to house a fire where the appliance manufacturer and/or flue box manufacturer indicate its suitability.

Generally, the following flue sizes are required, but individual manufacturer's instructions may specify differently.

Appliance Type	Flue Diameter (mm)
Radiant convector gas fire	125
Inset live fuel effect gas fire	125
Decorative fuel effect gas fire	175

(Purpose designed throats in DFE flue systems shall have no minor dimension less than 100 mm and a cross-sectional area not less than 240 cm^2.)

Catchment Spaces

Every gas fire that incorporates a flue spigot for dispersal of combustion products into the builder's opening or fireplace recess requires a void below the base of the flue spigot for the collection of debris.

Voids and Depths Below Gas Fire Flue Spigots

A gas fire should be fitted so that there is a void below the base of the appliance flue spigot for the collection of debris.

The volume of the void and the minimum depth is dependent upon whether:

- the chimney is of masonry or block construction
- the masonry chimney is unlined, or lined with clay or cement
- the chimney is new or unused, or previously used only with a gas appliance
- the chimney has previously been used with a solid fuel or oil-burning appliance.

The table below indicates these dimensions.

Table 1 Minimum void volumes and depths below gas fire flue spigots

Chimney type		Minimum depth (mm)	Minimum void volume dm^3 (litres)
Masonry	– unlined	250	12
	– lined (new or unused)	75	2
	– lined (previously used with other fuels)	250	12
Block	– new or unused	75	2
	– previously used with other fuels	250	12
Sheet Metal	– new or unused	75	2

Appliances with a direct flue connection must be connected to a lined flue. Freestanding appliances, e.g. heating stoves, installed in a recess with a register plate fitted above, should have their flue pipe projecting above the plate by at least 150 mm.

Closure Plate

Radiant convector gas fires shall always be fitted using a closure plate supplied with the fire unless otherwise specified in the manufacturer's instructions. These closure plates are of dimensions 660 mm x 460 mm incorporating an opening for the flue spigot and, where necessary, a ventilation or air relief opening to allow the correct flue flow rate when the appliance is in operation.

To prevent the entry of excess air into the flue, which would reduce the flue pull on the fire flue spigot, the closure plate shall be sealed to the fireplace wall or fire surround with adhesive tape or other sealing material capable of maintaining its seal up to a temperature of 100°C.

Any modification to a closure plate shall be in accordance with the manufacturer's instructions.

Where a closure plate is inadequate to fill the fireplace opening, or for other aesthetic reasons, an infill panel or surround shall be used with a suitable opening to which the closure plate may be secured.

Where the gas fire is brought forward from the fireplace opening (such as a low forward projection of a mantelshelf) a flue spigot extension may be used, up to 150 mm long. The flue spigot, or the spigot extension shall pass through the closure plate and extend through it for a minimum of 15 mm. There shall be a minimum of 50 mm clearance between the end of the spigot and the back of the fireplace opening or chairbrick.

When a closure plate is removed for inspection or servicing, the tape or other sealing material should be removed.

Inset fires to do not normally require a closure plate as the appliance body will usually be the closure plate. Each manufacturer will recommend the method of securing the fire to the builder's opening or flue box, and the sealing method to be used.

FIRE PRECAUTIONS

a) Hearths in builder's openings, fireplace recesses of flue boxes

The minimum depth of all hearths must not be less than 12 mm of fire resistant material. The manufacturer's instructions should, where necessary, give details of how to prevent the temperature rise on the underside of the hearth exceeding 80°C (normally on certain decorative fuel effect applications).

Hearths beneath live fuel effect and decorative fuel effect fires shall have the top surface of the hearth at least 50 mm above floor level, to discourage carpets or rugs being placed on top of the hearth. This can be achieved by having an upstand edge, or a fixed fender of 50 mm minimum height along the front and sides of the hearth.

The hearth shall extend at least 300 mm from:

- the back plane of a radiant convector fire

- in front of the naked flame or incandescent part of the firebed for inset live fuel effect or decorative fuel effect gas fires.

The hearth shall extend 150 mm beyond each side of the naked flame or incandescent radiant source, unless specified differently in the manufacturer's instructions.

b) Free-standing hearths and canopies for decorative fuel effect fires

The hearth must extend completely beneath and project outwards to at least 300 mm from all the sides of the naked flames or incandescent material.

Any part of the canopy or an uninsulated flue pipe which is within 1 m of the naked flame or incandescent material shall be a distance of not less than 300 mm from any combustible material. The canopy shall have no other openings than at its base and the flue outlet at the top.

c) Side wall protection

In the absence of any specific manufacturer's instructions, no fire shall be installed within 500 mm of any combustible wall when measured from the flame, incandescent radiant or material.

d) Shelf protection

A fire shall only be fitted below a shelf or other projection in accordance with the manufacturer's instructions.

e) Wall-mounted fires, e.g. raised builders openings

When wall-mounted, provided manufacturer's instructions state that no hearth is required beneath the appliance, then the distance from any flame or incandescent materials to any carpet or floor covering should be at least 225 mm. Where no carpet or floor covering exists, but the floor is of the type that is likely to be covered, then this distance should be increased to 300 mm.

f) User protection

Appliances may be fitted with an integral guard which should always be in place. Alternatively, fireguards complying with BS 8423 should be recommended to the user where young children, the elderly or infirm are likely to be present. Where an integral guard is not fitted as part of the appliance it is recommended that a 'tactile separator' in the form of a fender, kerb, horizontal bar or similar be placed between 50 mm and 1 m above floor level and in a similar position to the hearth edge (if fitted), i.e. 300 mm in front of and 150 mm beyond the edge of any naked flames or fire bed to prevent people inadvertently walking into or backing onto the fire.

Examples of methods of installing gas fires to their flue systems are shown in Figures 24–35.

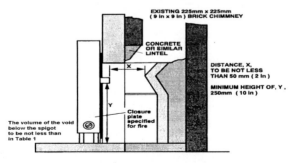

Figure 24 Method of installing a gas fire to an existing brick chimney – high mantelshelf

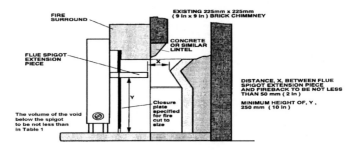

Figure 25 Method of installing a gas fire to an existing brick chimney – low and projecting mantelshelf

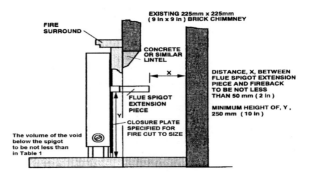

Figure 26 Method of installing a gas fire to a purpose-made hearth or surround

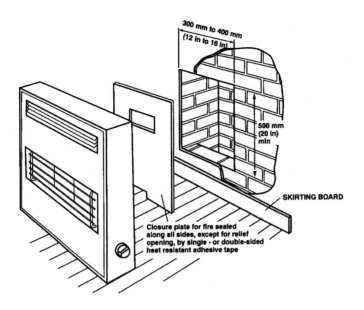

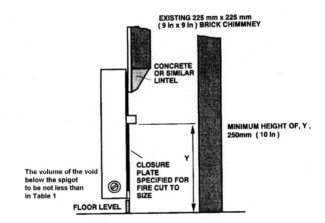

Figure 27 Method of wall-mounting a gas fire to a chimney breast

Measurement from base of flames to floor is normally 225 mm, or in accordance with the manufacturer's instructions.

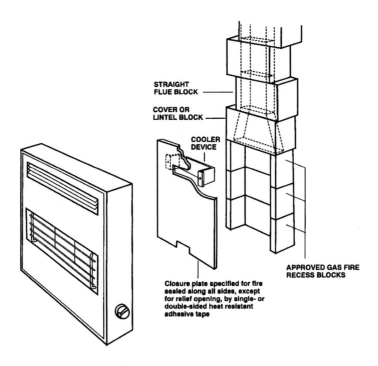

Figure 28 Method of installing a gas fire on the hearth of a pre-cast block construction flue

Cooler plate installation requirements must comply with manufacturer's instructions.

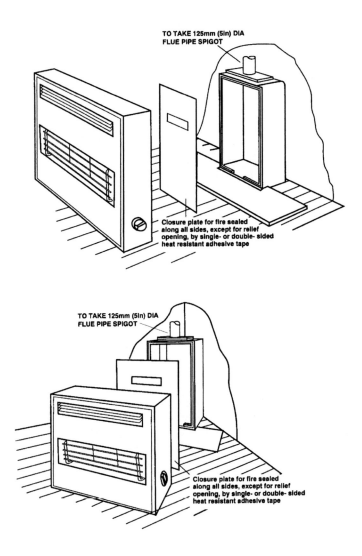

Figure 29 Method of installing a gas fire in conjunction with a flue box and flue pipe

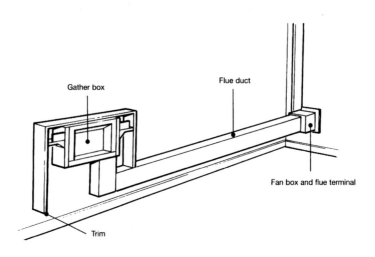

Figure 30 Installation of a side exit fanned-draught flue system

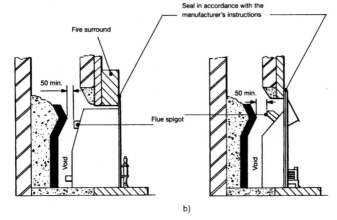

a)   b)

All dimensions are in millimetres

Note: The volume of the void below the appliance spigot to be in accordance with Table 1

Figure 31 Method of installing an inset live fuel effect gas fire to a masonry chimney (where chairbrick removal may not be necessary) using an existing solid fuel hearth

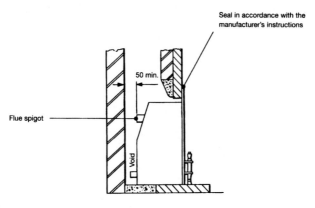

Dimension is in millimetres

Note: The volume of the void below the appliance spigot to be in accordance with Table 1

Figure 32 Method of installing an inset live fuel effect gas fire to a masonry chimney (where chairbrick removal may be necessary) using an existing solid fuel hearth

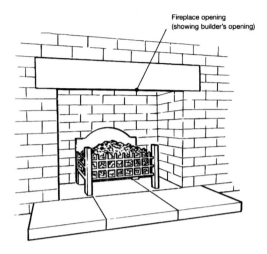

Note: Where a builder's opening does not have a smooth tapered transition (gather) to the flue a canopy fitted above the appliance tapering into the flue can be used to maintain the same effect.

Figure 33 Appliance installation in a typical builder's opening

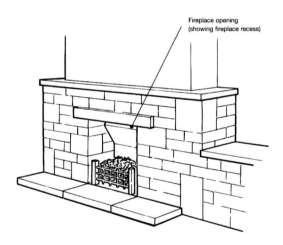

Note: A fireplace recess may contain either fireplace components complying BS 1251 and installed in accordance with BS 8303 or special fireplace components as supplied with the appliance or flue system.

Figure 34 Appliance installation in a typical fireplace recess

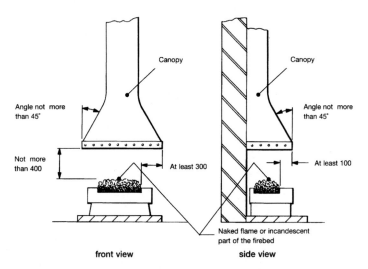

a) Associated independent canopy installation with appliance free-standing and positioned against a non-combustible wall

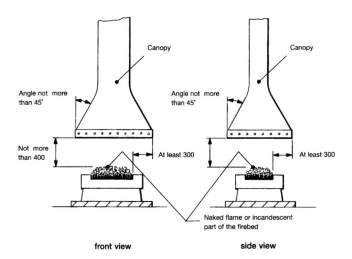

b) Associated independent canopy installation with appliance free-standing

Dimensions are in millimetres

Figure 35 Method of installing an appliance under an associated independent canopy

FLUE SIZING FOR DECORATIVE FUEL EFFECT (NATURAL DRAUGHT) FLUE SYSTEMS

When installing an appliance into an existing fireplace opening, it is not normally necessary to alter the flue size if the flue system has proven to work safely with a solid fuel open fire.

For a fireplace and flue system to work correctly there is a necessary relationship between the flue size, the flue height and the fireplace opening or canopy opening.

Fireplace openings up to 450 mm (18 in) wide by 600 mm (24 in) high will normally be successfully served by a flue having no dimension less than 175 mm (7 in) and a canopy that slopes up to the flue at no more than 45 degrees.

Where an appliance is installed in a fireplace opening or under a canopy that exceeds these dimensions (450 mm x 600 mm), use of the following flue sizing chart and diagrams showing fireplace/canopy openings will generally ensure satisfactory clearance of the flue products.

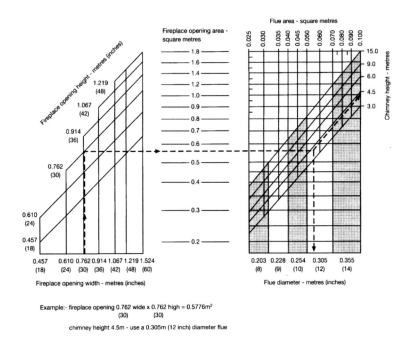

Example:- fireplace opening 0.762 wide x 0.762 high = 0.5776m²
                              (30)   (30)
         chimney height 4.5m - use a 0.305m (12 inch) diameter flue

Figure 36 Flue sizing chart for decorative fuel effect gas appliances with fireplace openings exceeding 0.457 m x 0.610 m

a) Area of fireplace opening is W x H

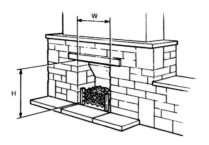

b) Area of fireplace opening is D x W

c) Area of fireplace opening is the area at base of canopy

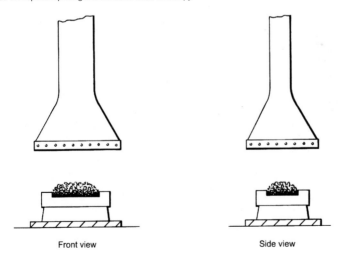

Front view   Side view

Note: This is applicable to all shapes of canopy base

Figure 37 Measurement of fireplace/canopy opening

FLUE TERMINATION FOR DECORATIVE FUEL EFFECT FIRES

Terminals and Chimney Pots

A terminal shall be fitted if the cross-sectional dimension of the flue is 170 mm or less across the axis of the outlet.

A terminal or chimney pot shall be fitted if this dimension exceeds 175 mm.

Proprietary terminals shall be in accordance with BS 715, BS EN 1858 or BS EN 1857.

Non-proprietary terminals shall be as follows:

- its total free area openings at least twice the nominal area of the flue
- outlet openings shall admit a 6 mm diameter ball but not a 16 mm diameter ball
- outlet openings shall be provided equally all round or on all sides of the terminal.

Chimney pots shall have no dimension across the axis of its outlet, or outlets, less than 175 mm.

Location

Terminal locations shall be in accordance with BS 5440-1 and shall not be sited within 600 mm of any other opening into a building, or the open end of a drainage ventilation pipe.

Bird Guards

Where there is a known problem that chimneys are used by nesting birds, squirrels or other wildlife a suitable guard or terminal shall be used which shall have a minor dimension of not more than 20 mm and be made from a material which is corrosion and weather resistant.

The chimney should be inspected and if necessary reinforced to ensure it will support the guard prior to fitting.

Figure 38 shows terminals likely to be suitable, and Figure 39 those likely to be unsuitable.

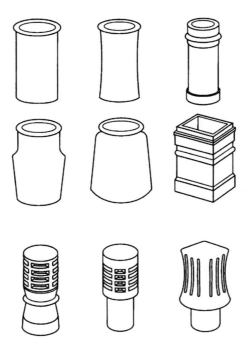

Figure 38 Terminations (likely to be suitable)

Outlets likely to be less than 175 mm across axis

Chimney pot inserts designed to cap off chimney but allow through ventilation

Figure 39 Terminations (likely to be unsuitable)

FLUE PROVING

PRE-INSTALLATION

It is essential to test any installation and flue to make sure it works correctly. The following paragraphs give essential information on how to achieve this, during installation or after servicing.

A preliminary inspection should be carried out to ascertain that:

- the flue is clean as far as can be seen and that any damper or restrictor plates are removed or permanently fixed in the open position
- the flue appears to be in good condition
- any terminal conforms to current standards
- where an appliance is already fitted, there is no sign of spillage (staining)
- the cross-sectional area of any flue is sufficient for the appliance being installed.

FLUE TESTING

Any chimney or flue system should be subjected to a flue flow test prior to the appliance being installed. This is to test, throughout its length, for:

- continuity
- integrity
- tightness.

The following procedure should be carried out:

- close all windows and doors in the room where the flue is to be tested
- check with a smoke match that there is a pull through the flue, just inside the builder's opening
- after confirming with a smoke match that smoke is drawn through into the flue, light a smoke pellet and observe that the flue is pulling the smoke away
- check outside that the smoke exits from one terminal and termination is suitable
- observe, where practicable, upstairs rooms, loft spaces, and other adjoining premises for signs of smoke, indicating a defective flue system.

This may require heat to be applied to the base of the flue to create a motive force through the flue system.

PREPARATION OF THE APPLIANCE

Having established that the flue is satisfactory for installing the gas fire, the gas fire will need to be prepared in accordance with the manufacturer's instructions:

- remove case and radiants
- remove transportation packaging and protection
- the flue spigot, in most cases, will need to be attached to the back of the fire.

To prevent against excessive flue pull through the fire a spigot restrictor may be required (supplied with the fire). The manufacturer's instructions will give guidance on when this plate should be fitted.

Before positioning on the hearth, the intended gas connection, method and route will require planning, as will any fittings or tube connected to the inlet of the fire.

The fire will need positioning so that the flue spigot locates through the closure plate. This may require, in the case of fireplaces with a low or projecting mantelshelf, the installation of a flue spigot extension.

Where a gas fire is being fitted to a pre-cast flue, a cooler plate may be required to prevent the overheating of this particular type of flue system. The plate is attached to the rear of the closure plate. Essential information regarding the fitting will be found in the manufacturer's instructions.

Any false chimney or one that has been lined must have the void between the flue and internal walls of the false chimney sealed. This can be achieved with a sealing plate or mineral wool.

GAS SUPPLY

It is a requirement of the Gas Safety (Installation and Use) Regulations that all flued appliances be fitted on a rigid or semi-rigid connection. Where an appliance is located in a flue box, the gas supply should only pass through the wall of the box as close as possible to the bottom and be adequately sealed with a non-setting sealant at its point of entry. An appliance isolation device is also required.

The accepted method for fixing a gas fire is shown in Figure 40.

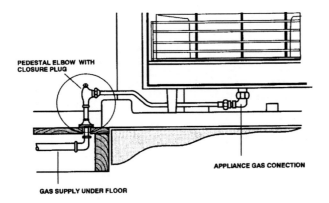

Figure 40 Typical gas supply connections to a gas fire

The connection is usually 8 mm copper (some may be brass or chrome finished). Springs are available to bend 8 mm copper. The 8 mm pipe should only be bent with a suitable copper bending spring. There are two types available, either internal or external.

All gas supplies should be installed in accordance with the minimum requirements of British Standards and Gas Safety Regulations. These include:

- pipe size in relation to the heat input of the appliance
- suitability of the pipe or tube for conveying gas
- jointing method
- fire precautions, i.e. the gas supply is not run in an obvious hot zone
- tightness
- support, corrosion protection and location.

For concealed connections any pipe passing through the builder's opening or buried in concrete must be adequately protected.

INSTALLING AND COMMISSIONING GAS FIRES CHECKLIST

- Preliminary examinations satisfactory i.e. data badge/installation instructions indicate suitability second-hand appliances visually non-defective.

- Not to be installed in prohibited locations.

- Ventilation requirements satisfactory.

- Chimney/flue systems satisfactory, i.e.
    - flue size and type
    - flue route
    - termination
    - catchment space (size and clear of debris).

- Flue flow test and inspection satisfactory (Ref. Gas Safety (Installation & Use) Regulations 1998 Regulation 27).

- Fire precautions satisfactory, i.e.
    - hearths
    - side wall protection
    - shelf protection
    - if wall-mounted height above floor satisfactory.

- Spigot restrictor fitted if necessary.

- Cooler device correctly located if necessary.

- Closure plate fitted correctly (radiant convector fire).

- Fire is secured and correctly sealed (inset gas fire).

- Isolation device correctly located.

- The appliance is correctly located (central fireplace) and stable.

- Gas supply correctly installed.

- The gas supply is gas-tight.

- The appliance is assembled correctly:
  - coal or log fuel effect or radiants correctly located.
- The appliance is commissioned in accordance with manufacturer's instructions and the current Gas Safety Regulations:
  - the appliance is purged of air
  - the working pressure is correct
  - the burner flame pictures, stability and ignition are correct
  - all safety control devices operate correctly (where fitted), i.e.
    - appliance isolation device
    - appliance integral gas taps
    - flame supervision devices
    - oxygen depletion systems (visual check for defective installation and flame pictures)
    - fan draught devices (fans and fan draught failure switches)
  - all user controls are operating satisfactorily (where fitted), i.e.
    - appliance integral gas taps
    - ignition devices
    - electric switches and bulbs or indicators
  - the spillage test is satisfactory.
- The safe operation and use of the appliance is explained and demonstrated.
- All manufacturer's instructions issued with the appliance are handed to the user.
- If the property is tenanted the landlord should be advised of their duties under the Gas Safety (Installation and Use) Regulations.

METHOD OF TESTING FOR SPILLAGE

Following installation or servicing the gas fire must be tested to ensure that the products of combustion are being evacuated to outside.

The requirements of BS 5440–1: 2000 and the manufacturer's instructions must be carried out on all open flued gas fires; a typical procedure is as follows:

- light the fire and allow the flue to warm for five minutes

- close all doors and windows in the room in which the appliance is installed

- with a smoke match in a suitable holder, test for spillage at the top of the radiant just inside the lip of the firebox canopy (some gas fires will have glass fronts, or it will be impossible to test above the radiants, in which case refer to the manufacturer's instructions)

- if there is any indication of spillage of products of combustion, the appliance should be left on for a further 10 minutes and the test repeated. If after this the fire fails the spillage test then, with the customer's permission, it should be disconnected, labelled and the action documented

- if there are no signs of smoke being pushed back into the room and it is drawn through the appliance into the flue, then the flue is working correctly.

Note: If there are other open flued appliances in the same room or adjoining room, then both flues will need testing, to make sure that one flue system does not affect the other. If there is an extractor fan in an adjacent room, the flue will need testing with the interconnecting doors open and any other doors and windows of that room closed.

INSTALLATION OF CONVECTOR HEATERS

BALANCED FLUE APPLIANCES

Location

Convector heaters can be installed in most rooms/areas of a property, the only considerations being:

- practicality of running the gas supply

- no installation in wet or damp environments, such as bathroom, for appliances with electrical connections

- flue route and termination
    - the wall that the flue passes through is suitably protected, both internally and externally, where fire protection is required
    - the flue is level
    - inlet and outlet ducts are cut to the correct length and sealed correctly
    - the terminal is protected with a guard where necessary

- fire precautions
    - proximity of furnishings in relation to the appliance must be taken into account before installation

- proximity of the electrical point, where required

- electrical cross-bonding

- the internal surface of the wall is suitable to allow installation.

FLUELESS SPACE HEATERS

Gas Supply

Flueless space heaters must be fitted by means of a rigid gas supply and connection. Flexible tubes are not permitted.

Appliance Fixing

The appliance must be secured to the wall or floor, taking into account fire precautions.

Ventilation

This appliance always requires additional purpose-provided ventilation. The ventilation must take the form of a purpose-provided airbrick or window vent taking air direct from outside. It is not permissible to ventilate these appliances from one room through another to outside, or from a ventilated void such as through a floor or loft space, unless ducted through the void to outside air.

MAINTENANCE OF SPACE HEATERS

The servicing of gas appliances is essential for their continued safe use and operating efficiency. The space heater is no exception.

The following should be carried out once a year.

OPEN FLUED HEATERS

Preliminary Inspection

- Check for general room ventilation.
- Visually check flue and termination.
- Visually check for signs of localised staining caused by spilled products of combustion.
- Check operation of controls, including flame supervision device if fitted.
- Check flame picture and ignition.
- Check for adequate purpose-provided ventilation where necessary.
- Check gas connection and stability of appliance.

Service

- Isolate appliance from gas supply and electricity supply.
- Remove case and clean internal surfaces of appliance.
- Disconnect appliance from gas supply and place safely to one side.
- Remove closure plate and remove any debris from catchment space (open flue gas fires).
- Check flue:
    - visually check that flue complies with BS 5440-1, and
        - is unobstructed and complete
        - services only one room or appliance
        - has the terminal correctly sited
        - any dampers are removed or fixed in the open position
    - a flue flow test is satisfactorily completed.

- Replace and seal closure plate (open flue gas fires).

- Clean burners and injectors of appliance.

- Ease and grease control tap as necessary.

- Clean and adjust any ignition device.

- Check heat exchanger and firebox, where applicable, for signs of cracking or damage.

- Check integrity of case or glass aperture seal.

- Check radiants, logs or imitation coals are clean and in good working condition (some appliances designed to use imitation logs or coals may require these components to be cleaned to remove any small deposits of soot).

- Replace, with customer's agreement, any defective parts.

- Reconnect appliance.

- Tightness test installation/appliance.

- Check burner pressure and where necessary gas rate input.

- Check that products of combustion are removed to outside air (spillage test).

- Discuss work carried out with customer.

ROOM SEALED HEATERS

Preliminary Inspection

- Visually check flue and termination.

- Check operation of controls, including flame supervision device if fitted.

- Check flame picture and ignition, where applicable.

- Check gas connection and stability of appliance.

Service

- Isolate appliance from gas supply and electricity supply.

- Remove case and clean internal surfaces of appliance.

- Clean burners and injectors of appliance.

- Ease and grease control tap as necessary.

- Clean and adjust any ignition device.

- Clean and inspect fan (fan balanced flue convectors).

- Check heat exchanger for signs of cracking or damage.

- Check integrity of case or glass aperture seal.

- Check radiants, logs or imitation coals are clean and in good working condition. (Some appliances designed to use imitation logs or coals may require these components to be cleaned to remove any small deposits of soot.)

- Replace, with customer's agreement, any defective parts.

- Reconnect appliance.

- Tightness test installation/appliance.

- Check burner pressure and gas rate input.

- Check that products of combustion are removed to outside air.

- Discuss work carried out with customer.

Tumble Dryers

CONTENTS

	Page
INTRODUCTION	1
The Gas Appliances Directive	1
Gas Appliances (Safety) Regulations 1995	1
European Standards	2
Regulations and Standards Affecting Installation and Maintenance	3
Health and Safety at Work etc. Act 1974 (HSW Act)	3
Management of Health and Safety at Work Regulations 1999 (MHSWR)	3
Reporting of Injuries, Diseases and Dangerous Occurrences Regulations 1995 (RIDDOR 95)	3
Building Regulations 2000 and Building (Scotland) Regulations 2004	3
British Standards	4
Manufacturer's Instructions	5
TUMBLE DRYERS	6
Component Parts	7
Case or Housing	7
Motor	7
Fan	7
Heater Assembly	7
Ignition and Flame Control Unit	7
Burners, Controls and Control Systems	8
Flame Protection Systems	8
Flame Rectification	8
Manual Controls	10
Automatic Controls	11
Summary of Operation	11
INSTALLATION OF TUMBLE DRYERS	13
Ventilation	13
Exhaust Vent	13
Location	14
Gas Supply	15
Electrical Connection	15
COMMISSIONING TUMBLE DRYERS CHECKLIST	16
SERVICING TUMBLE DRYERS	17

INTRODUCTION

This introduction contains an appraisal of some of the most important Regulations that determine both the design and manufacture of certain domestic appliances.

To reinforce this, there is reference to the Regulations and British Standards that domestic gas appliance installations must comply with. These Regulations and Standards also impose certain constraints on the subsequent maintenance of appliances and the pipework installation.

THE GAS APPLIANCES DIRECTIVE

Since 1985 the European Union has been pursuing a 'common approach' to tackle the problems associated with trade barriers between member states. For example, EU countries have different laws at present relating to product safety. Removing these barriers is at the heart of the Single European Market introduced in 1992.

The Gas Appliances Directive sets out legal requirements that in future will apply across the European Union. Member countries are required to amend their existing legislation, or to introduce new legislation that conforms with the requirements of the directive. The United Kingdom has implemented the Gas Appliances (Safety) Regulations to conform with the directive.

GAS APPLIANCES (SAFETY) REGULATIONS 1995

Until 1992, the safety (to consumers) of gas appliances sold in the United Kingdom has been covered by the Consumer Protection Act and specifically by the Gas Cooking (Safety) Regulations, and the Heating Appliances (Fireguards) Regulations. The Gas Appliances (Safety) Regulations 1995 introduced specific requirements.

There was a transitional period until 1996 in which gas appliances offered for sale in the United Kingdom were allowed to meet the old requirements.

The main provision of the new Regulations are:

a) **appliances must be safe**

b) **appliances must be tested**

c) **appliances must be quality guaranteed.**

This means that during the manufacturing process the manufacturer must operate a quality scheme of some type, such as BS 5750, to ensure that all appliances conform to the tested design. This scheme will be monitored by the 'Notified Bodies'.

d) **appliances must carry the CE mark**

All appliances that conform to provisions **a), b) and c)** will carry a CE mark issued by the 'Notified Bodies' (see Figure 1).

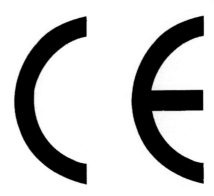

Figure 1 CE mark

The Regulations include detailed procedures for product conformity attestation by third party notified bodies, appointed by the Secretary of State.

All new gas appliances must have information included, covering safe installation, operation and maintenance.

EUROPEAN STANDARDS

European Standards are currently being compiled. For some appliances, where no European Standard is planned, the National Standards (in this country, British Standards) may be recognised. This, for example, will apply to the British type of gas fire.

REGULATIONS AND STANDARDS AFFECTING INSTALLATION AND MAINTENANCE

Health and Safety at Work etc. Act 1974 (HSW Act)

This Act applies to everyone concerned with work activities, ranging from employers, self-employed, and employees, to designers, suppliers and importers of materials for use at work, and people in control of premises. The duties apply both to individual people, and to corporations, companies, partnerships, local authorities etc. Employers have a duty to ensure, so far as is reasonably practicable, the health, safety and welfare at work of all employees, and not to expose people who are not their employees to risks to their health and safety.

Management of Health and Safety at Work Regulations 1999 (MHSWR)

These Regulations impose a duty on employers and self-employed persons to make suitable and sufficient assessment of risks to the health and safety of employees, and non-employees affected by their work. It also requires effective planning and review of protective measures, health surveillance, emergency procedures, information and training.

Reporting of Injuries, Diseases and Dangerous Occurrences Regulations 1995 (RIDDOR 95)

These Regulations require employers to report specified occupational injuries, diseases and dangerous occurrences (events) to the HSE. Certain gas incidents are reportable by suppliers of gas through fixed pipe distribution systems and/or LPG suppliers, and gas installers are required to report certain dangerous gas appliances to the HSE.

Building Regulations 2000 and Building (Scotland) Regulations 2004

These Regulations address the various aspects of building design and construction which include energy conservation and health and safety.

The Secretary of State has approved a number of documents under the Building Regulations 2000 as practical (non-mandatory) guidance to meeting the requirements under the Regulations.

Similar 'deemed to satisfy' guidance is provided in technical handbooks of the Building (Scotland) Regulations 2004.

The documents that particularly relate to gas work in domestic premises are:

- **Building Regulations 2000 (England and Wales)**

    Part   A   –   Structure

    Part   B   –   Fire Safety

    Part   F   –   Ventilation

    Part   G3   –   Hot Water Storage

    Part   J   –   Combustion Appliances and Fuel Storage Systems

    Part   L   –   Conservation of Fuel and Power

    Part   M   –   Access To and Use of Buildings

    Part   P   –   Electrical Safety

- **Building (Scotland) Regulations 2004**

    Section 1   –   Structure

    Section 2   –   Fire

    Section 3   –   Environment

    Section 4   –   Safety

    Section 6   –   Energy

BRITISH STANDARDS

British Standards' specifications are an invaluable guide to the installation of gas appliances. If followed, these standards will satisfy the requirements of current Regulations.

The following is a selection of some of the important British Standards Specifications relating to Domestic Gas Appliances, which give guidance on the minimum standard that appliance installations should comply with, to satisfy current Regulations:

BS 5546: 2000 — Specification for installation of gas hot water supplies for domestic purposes

BS 5588: (Domestic) Part 1 1990 — Fire precautions in the design, construction and use of buildings

BS 5864: 2004 — Installation and maintenance of gas fired ducted warm air heaters of rated input not exceeding 70 kW net (2nd and 3rd family gases)

BS 5871: Part 1 2005 — Gas fire, convector heaters and fire/back boilers (2nd and 3rd family gases)

BS 5871:	Part 2 2005	–	Inset fuel effect gas fires of a heat input not exceeding 15 kW (2nd and 3rd family gases)
BS 5871:	Part 3 2005	–	Decorative fuel effect gas appliances of a heat input not exceeding 20 kW (2nd and 3rd family gases)
BS 6700:	1997	–	Design, installation, testing and maintenance of water supplies for domestic purposes
BS 8423:	2002	–	Fire-guards for fires and heating appliances for domestic use - Specification
BS 6172:	2004	–	Installation and maintenance of domestic gas cooking appliances (2nd and 3rd family gases) – Specification
BS 6798:	2000	–	Specification for installation of gas fired boilers of rated input not exceeding 70 kW net
BS 6891:	2005	–	Installation of low-pressure gas pipework of up to 35 mm (R 1¼) in domestic premises (2nd family gases)
BS 5440:	Part 1 2000	–	Flues
BS 5440:	Part 2 2000	–	Air supply
BS 5482:	Part 1 2005	–	Code of practice for domestic butane and propane gas burning installations. Installations in permanent buildings, residential park homes and commercial premises up to 28 mm
BS 7624:	2004	–	Installation and maintenance of domestic direct gas fired tumble dryers up to 6 kW heat input (2nd and 3rd family gases) - Specification

MANUFACTURER'S INSTRUCTIONS

Manufacturer's instructions are important for the installation, commissioning, maintenance and use of any gas appliance. These instructions must be read and followed. Where manufacturer's instructions are not available, e.g. a used tumble dryer, they should be obtained before installation of the appliance begins.

After installation of the appliance or subsequent maintenance of it, the instructions must be returned to the consumer so that they may store them for future reference. This includes both user and installation/servicing instructions. (This is a requirement of the Gas Safety Regulations.)

TUMBLE DRYERS

Gas-fired tumble dryers have become very popular in the domestic market. They offer a saving in terms of running costs over their electric equivalents.

The domestic gas tumble dryer is a dual fuel appliance, the drying heat produced by gas and the tumbling action driven by electricity. Installation of appliances of up to 6 kW net heat input for both 2nd and 3rd family gases is specified in BS 7624: 2004.

These appliances are classified as flueless, although many are installed to an optional exhaust venting kit. Installation in bath and shower rooms is not permitted.

Installation within bedrooms or bed sitting rooms require a room volume of at least 7 m^3 per kW of appliance rated net heat input and the manufacturer's instructions must permit this location.

Air is taken from the room through the front of the appliance. A fully automatic burner heats a percentage of this air.

The heated air flows through a rear 'banjo' duct and into the perforated rear of the drum. Two inlet air temperature thermostats located on the rear banjo protect against an overheat condition occurring.

The moisture-laden air then passes from the drum through a removable lint filter to atmosphere. One of two exhaust thermostats monitors the exhaust air temperature, ready to switch off the burner in the event of a temperature rise caused by the clothes having dried in less time than the customer had set.

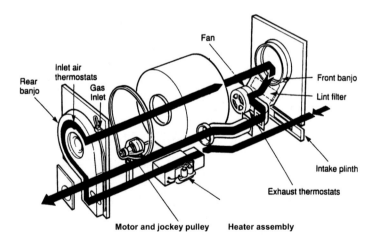

Figure 2 Major components and airflow

COMPONENT PARTS

Case or Housing

This is designed and manufactured to complement installation within fitted kitchen units. The dryer consists of a drum made of steel with either a painted or zinc coated finish, or a combination of both. The rear of the drum is supported by bearings to allow for rotation while the outer frontlip of the drum is supported by special moulded blocks.

Motor

Typically, a 240 volt motor turns the drum at approximately 60 revs/minute. The motor drives the drum by means of a belt – there may be some form of spring tensioner to take up slack in the belt.

Fan

Located normally at the front of the appliance and connected to the internal ducting system. The fan circulates air through the appliance.

Heater Assembly

The heater comprises of a tunnel-shaped, polished sheet steel flame tube assembly and a 5–8 bladed burner unit that fires sideways through the length of the flame tube. A 240 volt AC twin solenoid valve forms the main part of the gas train to the injector with a heat shield fitted between the flame tube and the solenoid valve.

Ignition and Flame Control Unit

Fully automatic ignition is controlled by an electronic flame control unit; high voltage spark ignition is provided through an electrode to light the burner. The flame supervision is controlled via the flame rectification probe.

The control unit monitors the appliance burning process. Any interruption in the burning cycle caused by fan failure, or loss of gas, will result in 'lock out'.

BURNERS, CONTROLS AND CONTROL SYSTEMS

FLAME PROTECTION SYSTEMS

Flame Rectification

This method of flame supervision superseded the more basic flame conductance system, which was prone to simulated d.c. flame signals. Condensation or a build up of carbon, due to flame chilling, can bridge the probe and burner. With a d.c. signal where electrons travel around the circuit in only one direction, a control unit can not distinguish the presence of a flame, from the bridging of the gap between the probe and the burner. However an a.c. signal can, due to the two directional flow to and from the control unit.

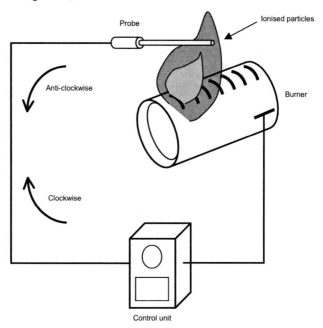

Figure 3 Flame rectification circuit

If we imagine the electrons from the control unit's signal, travelling in a clockwise direction, when the signal reaches the probe, the electrons are able to travel to the burner due to the ionised particles in the gas flame. If there was no flame present, the electrons are not supplied with sufficient pressure to jump the gap e.g. voltage/spark. The probe passing the electrons to the burner is very much like a shotgun firing pellets at a barn door (the burner has a much greater area). Therefore all of the electrons will travel the gap and be registered at the end of the clockwise journey back to the control unit.

When travelling back in the anti-clockwise direction, the electrons now try to pass the gap from the burner to the probe, this is now like shooting a cannon at a pencil, only some of the electrons are able to 'hit' the probe and travel back to the control unit. We now find a rectified signal recognised by the control unit as the presence of a flame.

The flame rectification system can distinguish various signals, for example:

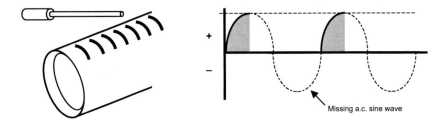

Figure 4 Open circuit

Figure 4, shows the signal read by the control unit where no flame or bridge is present. The electrons reach the end of the probe but have nowhere to go.

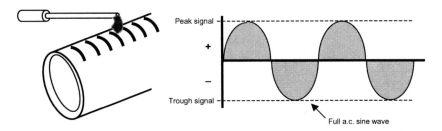

Figure 5 Closed circuit

Figure 5, shows the signal read by the control unit when the gap between the probe and burner, is bridged by conductive matter (condensation or carbon). The electrons can travel freely in both directions.

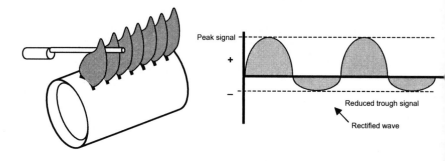

Figure 6 Rectified signal

Figure 6, shows the signal read by the control unit when the gap between the probe and burner, is bridged only by the flame, the shot gun and barn door effect now takes place, rectifying the signal.

A typical, burner head to probe ratio of 8:1 will allow a rectified signal to be produced. This ratio is easily surpassed for most atmospheric bar type burners.

Manual Controls

Door Switch

The appliance is deactivated electrically to prevent operation whenever the door is open. On closing the door, a switch contact is made, restoring the circuit through to the appliance controls.

Timer

A timer control can be preset from 0 up to 120 minutes. When the timer has been programmed to run the appliance, two microswitches make contact. One of the microswitches energises the gas circuit via the flame control unit, the other energises the fan motor, timer and drum motor.

The burner is programmed to shut down several minutes before the end of the drying period. This allows for cooling before the drum stops turning, thus minimising creasing of clothes.

Heat Switch

This allows the selection of either 'half' or 'full' heat. Effectively this selects either the 50°C or 60°C exhaust thermostat to control the maximum drying temperature.

Indicator Neon

This neon, indicating gas flow to the burner, illuminates whenever the burner solenoid valves are energised.

Automatic Controls

50°C or 60°C Exhaust Thermostats

These thermostats control the drying temperature to a maximum of 50°C or 60°C as selected by the position of the heat switch. They prevent damage to the clothes by ensuring they do not become too hot.

When the thermostats operate they de-energise the flame control unit and burner solenoids. When the thermostat has cooled down it will allow the flame control unit to re-energise the burner and cycle a number of times if the timer has not reached the cool down period.

110°C Inlet Thermostat

This thermostat protects the appliance against overheating caused by a restricted exhaust or overloading. As the thermostat is in a normally closed position, it will open to de-energise the flame control unit and solenoids while the fan continues to run until the timer stops. When the thermostat has cooled to approximately 95°C the burner will reignite if the timer is still running.

120°C Inlet Thermostat

This thermostat backs up the 110°C thermostat in the event that it fails to operate. When operated it will de-energise the burner solenoids and extinguish the neon indicator. Absence of the flame will cause the igniter electrode to spark for 10 seconds, but as the thermostat cannot cool down sufficiently in that time the control unit will go to a lock out condition. To clear this requires either the power to be turned off and on again or the door opened and closed.

Summary of Operation

With the door switch closed and the timer set, the drum motor, fan assembly, timer motor and flame control unit are energised. Power to the flame control unit is routed through the 50°C or 60°C exhaust thermostat in accordance with the position of the heat selector switch.

After an air purging period of approximately 9 seconds, the solenoid valves and spark generator are energised and the neon indicator illuminated. If the flame electrode does not detect a flame within 10 seconds of sparking it will cause the unit to enter a lock out condition, however, the fan will continue to run. In order to reset the control unit either the power should be turned off and on, or the door opened and closed.

During operation the selected exhaust thermostat will protect the clothes from overdrying by de-energising and re-energising the gas valves via the control unit.

Twelve minutes before the drying period ends, the timer will de-energise the flame unit (to provide a cooling period) but the fan and drum will continue rotating until the timer stops.

Any restriction to the flow of heated air is protected by the 110°C and/or the 120°C thermostats.

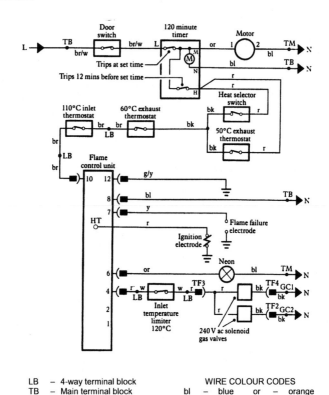

LB	– 4-way terminal block		WIRE COLOUR CODES		
TB	– Main terminal block	bl – blue		or – orange
TF	– Tab/faston housing	bk – black		r – red
GC	– Flame control unit terminals	br – brown		w – white
TM	– Timer	g – green		y – yellow
HT	– High tension				

Figure 7 A typical function flow diagram showing tumble dryer controls

INSTALLATION OF TUMBLE DRYERS

VENTILATION

Ventilation is an important consideration. A suitable room will need an openable window, or equivalent in accordance with Building Regulations.

Any room with a volume of less than 3.7 m^3/kW, will need additional ventilation of 100 cm^2.

The consumer should be advised of the need to open the window while using this type of appliance, to prevent excessive levels of condensation.

The ventilation for a gas tumble dryer must be calculated in relation to other appliances within the same room. For example, ventilation in a kitchen with a cooker and an open flued boiler will need calculating in accordance with the recommendations of BS 5440-2.

It is also important to test that exhaust-vented tumble dryers are not having an adverse effect on the flue of any open flued appliance installed in the same room.

Exhaust Vent

The main reason for installing an exhaust vent is to overcome the considerable amount of water vapour being produced, mainly by the drying process. Tumble dryers that are installed under a working surface **must** be fitted with an exhaust vent kit and it is recommended that **all** tumble dryer installations are vented to outside.

There are two basic methods of installing a permanent exhaust vent:

- through a wall (Figure 8)

- through a single-glazed window (Figure 9).

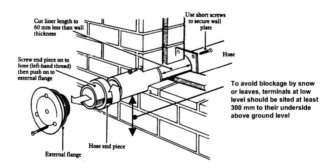

Figure 8 Wall installation method for a tumble dryer exhaust vent

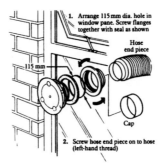

Figure 9 Window installation method for a tumble dryer exhaust vent

It should be noted that where the exhaust hose is hung directly out of the window the tumble dryer should not exceed 3 kW net heat input.

LOCATION

The location of the appliance should be carefully considered to ensure that there is adequate room. The minimum dimensions for any individual appliance must be checked with the manufacturer's instructions. In the case of freestanding installations, the floor or base needs to be level.

If the appliance is likely to be moved regularly, or is to be installed on top of a washing machine using a stacking kit, then a restraining cable must be fitted to the flexible gas connection to prevent any damage to the supply through movement of the appliance (Figure 10).

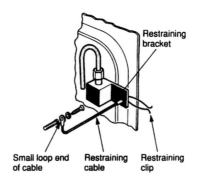

Figure 10 A typical restraining kit

GAS SUPPLY

Although tumble dryers have a relatively low gas rate, the size of the gas supply still needs to be considered in relation to other appliances fed from the same supply.

Gas connection to the appliance is normally by cooker flexible tube and bayonet connection.

ELECTRICAL CONNECTION

The dryer should be connected using a suitable 13 amp plug correctly fused in accordance with the manufacturer's instructions.

An unswitched electrical socket outlet is recommended to encourage removal of the electric plug during servicing.

COMMISSIONING TUMBLE DRYERS CHECKLIST

1. Test the supply for gas tightness.
2. Purge supply.
3. Connect a gauge to the appliance test point.
4. Set the heat switch to high with the timer off.
5. Ensure that the drum is empty and that the lint filter is fitted before closing the door.
6. Turn on the gas and electrical supply.
7. Set the timer to 30 minutes and check that the drum rotates and burner ignites.
8. Ensure that the indicator neon has illuminated.

    If the burner fails to light, or to remain alight, turn the timer off and on for 30 second periods until the air is purged and the burner is on continuously.

9. Ensure that the pressure is 20 mbar or as recommended by the manufacturer's instructions.
10. Remove the gauge and check for leaks.
11. Check the gas rate if necessary.
12. Restart the dryer and disconnect the gas supply hose.
13. Ensure the gas solenoid valves are de-energised and burner extinguished.
14. Reconnect hose after 30 seconds, there should be no ignition.
15. Turn the dryer off, then on again to check for reignition.
16. Check for spillage of combustion products from any open flued appliances installed in the same (or adjacent) room if exhaust is ducted to outside air, ensuring any additional extract fans, etc. are also in operation.
17. Advise customer on use of appliance.
18. Ensure manufacturer's instructions are left with customer.

SERVICING TUMBLE DRYERS

1. Isolate gas and electricity supplies.
2. Remove filter and clean.
3. Ensure air intake grill and intake gauge are clean.
4. Remove burner from heater assembly.
5. Clean burners and injectors and reassemble.
6. Ensure vent hose and fan are clean and in good condition.
7. Ensure drum rotates freely and is free from lint debris.
8. Reassemble and check for leaks.
9. Check operation of all controls/components including flame supervision device.
10. Light appliance and check burner pressure and/or gas rate are in accordance with manufacturer's instructions.
11. Check for spillage of combustion products from any open flued appliances installed in the same (or adjacent) room if exhaust is ducted to outside air.
12. Emphasize to customer the importance of cleaning the lint filter after every drying cycle.

Instantaneous Water Heaters

CONTENTS

	Page
INTRODUCTION	1
The Gas Appliances Directive	1
Gas Appliances (Safety) Regulations 1995	1
European Standards	2
Regulations and Standards Affecting Installation and Maintenance	3
Health and Safety at Work etc. Act 1974 (HSW Act)	3
Management of Health and Safety at Work Regulations 1999 (MHSWR)	3
Reporting of Injuries, Diseases and Dangerous Occurrences Regulations 1995 (RIDDOR 95)	3
Building Regulations 2000 and Building (Scotland) Regulations 2004	3
British Standards	4
Manufacturer's Instructions	5
INSTANTANEOUS WATER HEATERS	6
Sink Water Heaters	6
Multi-point Water Heaters	7
Construction and Operation of Water Heaters	7
Heating Body	8
Water Diaphragm Housing	9
Slow Ignition Device	10
Non-thermostatic Temperature Control	12
Thermostatic Control	13
Temperature Selector	13
Water Throttle and Water Governor	14
Gas Valve Housing	16
Volumetric Governor	16
Ignition Source	17
Flame Supervision Device	17
Thermo-electric	17
Oxygen Depletion System (ODS) – Atmosphere Sensing Device (ASD)	20
Flue Safety Devices	23
Flame Rectification	24
Burner	26
Location of Components	26

(continued overleaf)

	Page
INSTALLATION OF INSTANTANEOUS WATER HEATERS	30
Location Guidelines	31
Small Single Point Instantaneous Water Heaters	31
Instantaneous Multi-point Water Heaters	32
Mains Fed Systems	32
Cistern Fed Systems	32
Flues and Ventilation	33
Bathrooms and Shower Rooms	33
Living Rooms, Kitchens, Utility Rooms, Halls and Passageways	33
Cloakrooms and Toilets	34
Compartments and Cupboards	34
Under Stairs Cupboards	34
Bedrooms and Bedsitting Rooms	35
Roof Space Installations	35
Water Supply	36
Materials	36
Pipe Sizes	36
Pipe Supports and Fixings	37
Water Pipework – General Guidelines	37
Back Siphonage	39
Frost Protection	39
Scale and Corrosion	39
Gas Supply	41
Protection and Jointing	41
Temporary Continuity Bonds	42
Electrical Supply (Where Applicable)	42
INSTALLATION OF SHOWER UNITS	44
Gravity Showers	44
Mains Pressure Showers	44
Pumped Showers	45
Shower Units – General Guidelines	45
INSTALLATION OF WASHING MACHINES AND DISHWASHERS (HOT FILL)	47
COMMISSIONING CHECKLIST FOR INSTANTANEOUS WATER HEATERS	48
MAINTENANCE CHECKLIST FOR INSTANTANEOUS WATER HEATERS	49
Pre-service Checks	49
Full Service	49
FAULT-FINDING CHECKLIST FOR INSTANTANEOUS WATER HEATERS	51

INTRODUCTION

This introduction contains an appraisal of some of the most important Regulations that determine both the design and manufacture of certain domestic appliances.

To reinforce this, there is reference to the Regulations and British Standards that domestic gas appliance installations must comply with. These Regulations and Standards also impose certain constraints on the subsequent maintenance of appliances and the pipework installation.

THE GAS APPLIANCES DIRECTIVE

Since 1985 the European Union has been pursuing a 'common approach' to tackle the problems associated with trade barriers between member states. For example, EU countries have different laws at present relating to product safety. Removing these barriers is at the heart of the Single European Market introduced in 1992.

The Gas Appliances Directive sets out legal requirements that in future will apply across the European Union. Member countries are required to amend their existing legislation, or to introduce new legislation that conforms with the requirements of the directive. The United Kingdom has implemented the Gas Appliances (Safety) Regulations to conform with the directive.

GAS APPLIANCES (SAFETY) REGULATIONS 1995

Until 1992, the safety (to consumers) of gas appliances sold in the United Kingdom has been covered by the Consumer Protection Act and specifically by the Gas Cooking (Safety) Regulations, and the Heating Appliances (Fireguards) Regulations. The Gas Appliances (Safety) Regulations 1995 introduced specific requirements.

There was a transitional period until 1996 in which gas appliances offered for sale in the United Kingdom were allowed to meet the old requirements.

The main provision of the new Regulations are:

a) **appliances must be safe**

b) **appliances must be tested**

c) **appliances must be quality guaranteed.**

This means that during the manufacturing process the manufacturer must operate a quality scheme of some type, such as BS 5750, to ensure that all appliances conform to the tested design. This scheme will be monitored by the 'Notified Bodies'.

d) appliances must carry the CE mark

All appliances that conform to provisions **a), b) and c)** will carry a CE mark issued by the 'Notified Bodies' (see Figure 1).

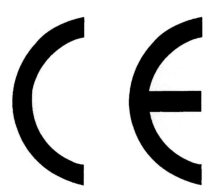

Figure 1 CE mark

The Regulations include detailed procedures for product conformity attestation by third party notified bodies, appointed by the Secretary of State.

All new gas appliances must have information included, covering safe installation, operation and maintenance.

EUROPEAN STANDARDS

European Standards are currently being compiled. For some appliances, where no European Standard is planned, the National Standards (in this country, British Standards) may be recognised. This, for example, will apply to the British type of gas fire.

REGULATIONS AND STANDARDS AFFECTING INSTALLATION AND MAINTENANCE

Health and Safety at Work etc. Act 1974 (HSW Act)

This Act applies to everyone concerned with work activities, ranging from employers, self-employed, and employees, to designers, suppliers and importers of materials for use at work, and people in control of premises. The duties apply both to individual people, and to corporations, companies, partnerships, local authorities etc. Employers have a duty to ensure, so far as is reasonably practicable, the health, safety and welfare at work of all employees, and not to expose people who are not their employees to risks to their health and safety.

Management of Health and Safety at Work Regulations 1999 (MHSWR)

These Regulations impose a duty on employers and self-employed persons to make suitable and sufficient assessment of risks to the health and safety of employees, and non-employees affected by their work. It also requires effective planning and review of protective measures, health surveillance, emergency procedures, information and training.

Reporting of Injuries, Diseases and Dangerous Occurrences Regulations 1995 (RIDDOR 95)

These Regulations require employers to report specified occupational injuries, diseases and dangerous occurrences (events) to the HSE. Certain gas incidents are reportable by suppliers of gas through fixed pipe distribution systems and/or LPG suppliers, and gas installers are required to report certain dangerous gas appliances to the HSE.

Building Regulations 2000 and Building (Scotland) Regulations 2004

These Regulations address the various aspects of building design and construction which include energy conservation and health and safety.

The Secretary of State has approved a number of documents under the Building Regulations 2000 as practical (non-mandatory) guidance to meeting the requirements under the Regulations.

Similar 'deemed to satisfy' guidance is provided in technical handbooks of the Building (Scotland) Regulations 2004.

The documents that particularly relate to gas work in domestic premises are:

- **Building Regulations 2000 (England and Wales)**

    | Part | A | – | Structure |
    | Part | B | – | Fire Safety |
    | Part | F | – | Ventilation |
    | Part | G3 | – | Hot Water Storage |
    | Part | J | – | Combustion Appliances and Fuel Storage Systems |
    | Part | L | – | Conservation of Fuel and Power |
    | Part | M | – | Access To and Use of Buildings |
    | Part | P | – | Electrical Safety |

- **Building (Scotland) Regulations 2004**

    Section 1 – Structure
    Section 2 – Fire
    Section 3 – Environment
    Section 4 – Safety
    Section 6 – Energy

BRITISH STANDARDS

British Standards' specifications are an invaluable guide to the installation of gas appliances. If followed, these standards will satisfy the requirements of current Regulations.

The following is a selection of some of the important British Standards Specifications relating to Domestic Gas Appliances, which give guidance on the minimum standard that appliance installations should comply with, to satisfy current Regulations:

BS 5546: 2000	–	Specification for installation of gas hot water supplies for domestic purposes
BS 5588: (Domestic) Part 1 1990	–	Fire precautions in the design, construction and use of buildings
BS 5864: 2004	–	Installation and maintenance of gas fired ducted warm air heaters of rated input not exceeding 70 kW net (2nd and 3rd family gases)
BS 5871: Part 1 2005	–	Gas fire, convector heaters and fire/back boilers (2nd and 3rd family gases)

BS 5871:	Part 2 2005	–	Inset fuel effect gas fires of a heat input not exceeding 15 kW (2nd and 3rd family gases)
BS 5871:	Part 3 2005	–	Decorative fuel effect gas appliances of a heat input not exceeding 20 kW (2nd and 3rd family gases)
BS 6700:	1997	–	Design, installation, testing and maintenance of water supplies for domestic purposes
BS 8423:	2002	–	Fire-guards for fires and heating appliances for domestic use - Specification
BS 6172:	2004	–	Installation and maintenance of domestic gas cooking appliances (2nd and 3rd family gases) – Specification
BS 6798:	2000	–	Specification for installation of gas fired boilers of rated input not exceeding 70 kW net
BS 6891:	2005	–	Installation of low-pressure gas pipework of up to 35 mm (R 1¼) in domestic premises (2nd family gases)
BS 5440:	Part 1 2000	–	Flues
BS 5440:	Part 2 2000	–	Air supply
BS 5482:	Part 1 2005	–	Code of practice for domestic butane and propane gas burning installations. Installations in permanent buildings, residential park homes and commercial premises up to 28 mm
BS 7624:	2004	–	Installation and maintenance of domestic direct gas fired tumble dryers up to 6 kW heat input (2nd and 3rd family gases) - Specification

MANUFACTURER'S INSTRUCTIONS

Manufacturer's instructions are important for the installation, commissioning, maintenance and use of any gas appliance. These instructions must be read and followed.

After installation of the appliance or subsequent maintenance of it, the instructions must be returned to the consumer so that they may store them for future reference. This includes both user and installation/servicing instructions. (This is a requirement of the Gas Safety Regulations.)

INSTANTANEOUS WATER HEATERS

Instantaneous gas-fired water heaters are designed to provide hot water when a draw-off tap is opened on the outlet supply of the appliance. To provide water at the temperature required and to avoid any danger of the water boiling and producing steam, normally the heaters are designed to raise the incoming water temperature through a maximum of 55°C (100°F).

In addition to its capacity for supplying hot water to taps throughout the customer's home, the instantaneous water heater may also supply a wide variety of outlets such as:

- automatic washing machines
- dishwashers
- showers.

To allow for the satisfactory operation of the instantaneous water heater, the installer must ensure that additional outlets to be supplied are fully compatible with the water heater being installed and that water flow to and from the heater is adequate. The manufacturer's instructions should be consulted for specific requirements and limitations.

There have been many designs of instantaneous water heater since its introduction onto the market in the early 1930s. Today they fall into two main categories:

- sink water heaters
- multi-point water heaters.

SINK WATER HEATERS

These types of water heater can be used with a swivel spout to deliver hot water to one or two sinks in close proximity to the appliance, or with a pipework system to a draw-off tap or shower. A typical heat input is about 11 kW (38,000 Btu/h) with a water flow rate of approximately 2.2 litres/minute raised through 55°C.

Although they are normally fitted without the need of a flue system, if the continuous use is to exceed 5 minutes when supplying a shower, or when a hot draw-off is not located in the same room or space as the heater, then a flue must be fitted.

A warning label should be displayed on a flueless appliance restricting the continuous use to a maximum of 5 minutes. Manufacturer's instructions should be consulted regarding the need for permanent ventilation.

Figure 2 Sink water heater warning label

MULTI-POINT WATER HEATERS

This appliance can supply hot water to a wide range of draw-offs connected to the outlet of the heater. A typical heat input of one of these appliances is approximately 29 kW (100,000 Btu/h) and they have a water flow rate that will normally only supply one outlet at a time (approximately 5.8 litres/minute raised 55°C, or 9 litres/minute raised 36°C).

Where more than one outlet is to be used simultaneously a storage system may be more appropriate. Instantaneous water heaters may be connected to a mains water supply or to a suitable cold feed cistern subject to the model of appliance being installed.

All multi-point water heaters must be connected to a flue to remove the products of combustion. For safety reasons the use of room sealed appliances is promoted by Government Regulations and British Standards. Although common practice prior to 1984, current regulations prevent the installation of open flued appliances in bathrooms and shower rooms.

CONSTRUCTION AND OPERATION OF WATER HEATERS

When a water draw-off tap is opened on the outlet side of the appliance, cold water enters the heating body and simultaneously opens a gas valve. As the water passes around the heat exchanger it is rapidly heated and delivered to the draw-off. When the outlet is closed the water stops flowing and the gas valve closes. Although the gas rate of these appliances is relatively high to produce an intense 'heat exchange' process there will always be a delay in hot water being available at the draw-off from initial start up. This delay will be dependent on the length and route of the pipework between the heater and the draw-off taps, as with any hot water system.

The main components of an instantaneous water heater are:

- a heating body comprising of the heat exchanger and combustion chamber
- a water diaphragm housing with water filter, throttle, thermostat, slow ignition device and temperature selector
- a gas valve housing
- an ignition source
- a flame supervision device
- a burner.

Balanced flue multi-points also comprise a balanced flue duct and terminal.

Heating Body

A typical heating body, as shown below (Figure 3) comprises a combustion chamber, into which the gas burner fires, and a heat exchanger which allows the cold water to pass above the heat source before being delivered to the outlet draw-off tap. Heating bodies are normally made of copper, covered by a lead and tin coating.

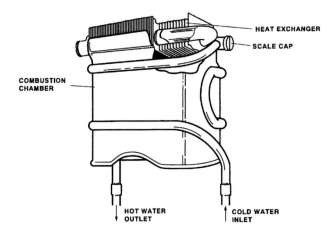

Figure 3 Heating body

The heat exchanger is designed so that the cold water passes several times through pipes fitted with banks of fins to maximise the heat gain from the burning gas.

Heat exchangers used in hard water areas are prone to the waterways becoming clogged with scale. On some heaters the component parts can be removed and descaled, or replaced.

Water Diaphragm Housing

On all heaters currently available, the water passing through the water housing 'lifts' a diaphragm onto which is connected a valve spindle and gas valve (Figure 4). This device is normally referred to as a 'differential pressure valve'.

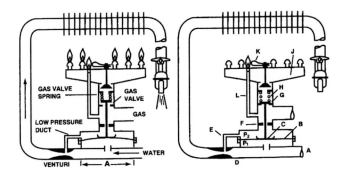

A: Diaphragm housing; **B**: Rubber diaphragm; **C**: Bearing plate; **D**: Venturi;
E: Low pressure; **F**: Stuffing box or gland; **G**: Gas valve; **H**: Gas valve spring;
J: Burner; **K**: Flame supervision device (bimetallic); **L**: Pilot supply.

Figure 4 Differential pressure valve

A device, called a 'venturi', fitted in the outlet side of the water diaphragm housing, increases the water velocity and reduces the water pressure as it passes through the narrow opening of the venturi, when water is flowing. The lower pressure (P_2 on Figure 4), is transmitted above the diaphragm, while the underside is subjected to the higher pressure (P_1 on Figure 4).

The difference in pressure above and below the diaphragm is sufficient to force the water upwards providing the flow rate is above the minimum rate set by the manufacturer. The gas valve, connected to the diaphragm, is lifted off its seating and gas is allowed to pass to the main burner where it is ignited and subsequently heats the water. If insufficient water is flowing through the machine the diaphragm will not lift and no gas will pass to the burner.

When the outlet is turned off and water stops flowing, the pressure above and below the diaphragm (Figure 4, P_1 and P_2) equalises and a spring pushes the gas valve back to its seating, cutting the gas supply to the burner.

Diaphragms can be made of rubber, neoprene or plastic and although they require little or no maintenance they can become distorted, which may result in the gas valve not completely opening.

Because of the relatively high gas rate of instantaneous water heaters the gas valve is designed to lift slowly, or in stages, to allow smooth ignition of the gas.

Slow Ignition Device

To ensure that the main gas burner ignites safely and quietly, most differential pressure valves incorporate a 'slow ignition device'. The device is designed to make the gas valve open slowly ensuring quiet, non-explosive ignition and close quickly preventing any water remaining in the heater overheating. They are housed in either the low or high pressure ducts to the diaphragm and can be either a fixed or adjustable type as illustrated below.

Fixed Type

This can be found in the high pressure inlet to the diaphragm (see Figure 5). It comprises a small disc with a hole in it free to float up and down in a cylindrical, brass housing. The rate at which the diaphragm lifts is controlled by the disc. When water flows through the heater the disc is forced upwards against its seating allowing water only to pass through the small centre hole. This allows the valve to lift slowly, turning the main gas fully on, after ignition. When water stops flowing the diaphragm moves downwards and the disc falls to rest on the stops. The gas valve closes quickly as water passes through the centre hole and around the edges of the disc.

Adjustable Type

This can be found in the low pressure inlet to the diaphragm (see Figure 6) and comprises a loose ball instead of a disc. Screwing the device in or out it regulates the amount of water which can pass by the ball, adjusting the rate of lift of the diaphragm. The ball follows a similar principle to the disc, where it restricts the waterways when the gas valve is opening and opens them to allow the gas valve to shut quickly.

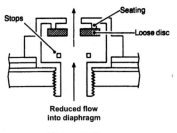

a) Valve opening

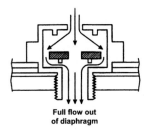

b) Valve closing

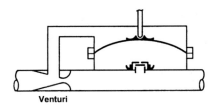

c) Disc type slow ignition device in automatic valve

Figure 5 Fixed (disc type) slow ignition device

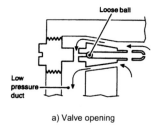

a) Valve opening

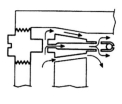

b) Valve closing

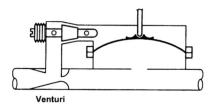

c) Adjustable slow ignition device in automatic valve

Figure 6 Adjustable (ball type) slow ignition device

Non-thermostatic Temperature Control

Once the gas valve has opened fully the temperature of the water at the outlets can be adjusted by partially closing the draw-off and thus reducing the amount of water passing through the heat exchanger. However, if the draw-off is closed down too far, the flow rate will fall below the design minimum, the gas valve will close and cold water will be delivered to the draw-off. On some heaters a simple temperature selector is fitted, which adjusts the water rate through the machine.

Thermostatic Control

A method of controlling the flow rate and outlet water temperature can be achieved by utilising a vapour-pressure thermostat. (Figure 7)

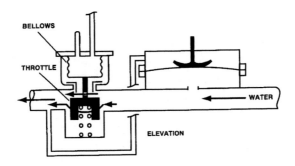

Figure 7 Thermostatic valve

A temperature sensing probe is located in the heat exchanger of the heater. This probe is connected by a small capillary tube to a set of bellows and a throttle located in the water section housing.

When a draw-off tap is opened and water begins to flow, the throttle is partially closed and only allows the minimum flow rate to pass. A difference in pressure above and below the diaphragm is achieved which controls gas flow via the main gas valve. As the water heats up the bellows expand and open the throttle, thus increasing the flow rate of the water and lowering its temperature. The temperature is usually set at 60°C (140°F) for this type of control on a multi-point heater.

Temperature Selector

On heaters without a thermostat, the hot water temperature can be varied by using the hot water tap to control the flow rate. With the tap partially closed, the water temperature will increase. However, if the tap is closed to less than the minimum flow required to operate the valve, the gas will shut off and the water will run cold. A temperature selector avoids this problem.

The temperature selector, illustrated in Figure 8, consists of a manually operated, screw-in throttle, situated in a bypass around the venturi.

When the throttle is fully screwed in, all the water passes through the venturi. This causes the automatic valve to operate and gives the highest temperature rise.

As the throttle is screwed out, an increasing amount of water bypasses the venturi and the hot water temperature is reduced.

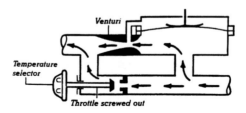

Figure 8 Temperature selector

Water Throttle and Water Governor

The amount of water passing through the machine can be set by adjusting a throttle screw, located prior to the diaphragm housing. A filter, which should be cleaned out periodically, may also be incorporated within this device (Figure 9).

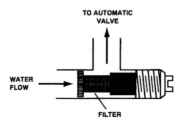

Figure 9 Adjustable water throttle

As the pressure from mains fed systems can fluctuate a water governor may be used on the inlet, or as an integral part of the water diaphragm housing.

In Figure 10 an independent water governor is shown which gives a constant outlet water pressure. If the outlet has a fixed area/diameter it will also give a constant volume, or rate of flow of water.

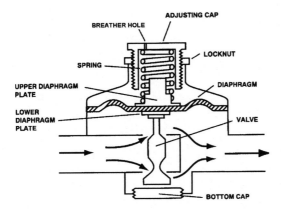

Figure 10 Water pressure governor

Figure 11 shows a water volume governor that is contained within the water diaphragm housing. When water passes through the heater the diaphragm rises and the governor valve is moved towards its seating with the pressure of the spring. Any fluctuation in inlet water pressure will move the position of the diaphragm and subsequently the governor valve to adjust the outlet pressure accordingly.

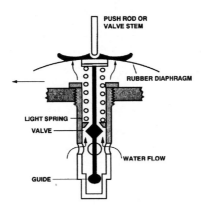

Figure 11 Water volume governor

GAS VALVE HOUSING

Volumetric Governor

A volumetric governor (Figure 12) may be fitted where the amount of gas being burnt is fixed and constant. This type of governor has the advantage over constant pressure governors; maintaining the correct gas rate irrespective of inlet pressure fluctuation and minor alterations in the size of burner orifices due to expansion.

Further information on constant pressure governors can be found in the ConstructionSkills publication *Gas Safety*.

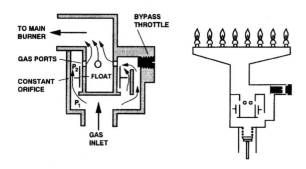

Figure 12 Volumetric governor

The volumetric governor consists of a lightweight 'float' which slides freely over a vertical tube, while maintaining a piston fit. The tube has a number of holes through which gas passes to the burner. These holes are covered as the float rises and uncovered when it falls. A fixed orifice, shown in the diagram between the float mechanism and the outer wall, creates a pressure difference which supports the float in a predetermined position when the correct volume of gas is passing. If the inlet pressure increases, the rate of flow through the fixed orifice rises and the pressure difference across the float also increases. The float therefore rises and partially closes the holes in the tube until the correct volume of gas is passing to the burner.

If the inlet pressure falls, the rate of flow reduces and the pressure difference allows the float to fall, exposing a greater area of the holes and thus increasing the flow rate to the volume required.

Ignition Source

The majority of instantaneous water heaters rely on a permanent pilot light to ignite the main gas. This pilot light may be lit manually, or more commonly by a piezo-electric spark operated by a separate push button.

Flame Supervision Device

All instantaneous water heaters are fitted with protection to prevent unlit gas being present in the combustion chamber if the pilot light goes out.

Flame supervision devices must detect the presence of a flame in order to maintain the gas supply to the burner, and shut off that supply if the flame is not present. Current British Standards specify the maximum shutdown time for flame supervision devices.

Thermo-electric

In its simplest form, the thermo-electric device (thermocouple) is a loop of two dissimilar metals joined together at one end, with the other ends connected to an electro-magnet. When the joint or junction is heated, a small voltage is produced (see figure 13).

The voltage produced is dependent on the temperature and the metals used. Generally thermocouples used as flame supervision devices utilise a chrome-nickel alloy and copper. The output voltage produced for these metals is between 15 to 30 mv. When the joint is heated by a pilot flame, the voltage energises the magnet thus holding the armature to it in a spring-operated gas valve and allowing gas to flow to the main burner.

Should the pilot be extinguished, the thermocouple would cool down and stop producing a voltage, thus allowing the spring to close the valve.

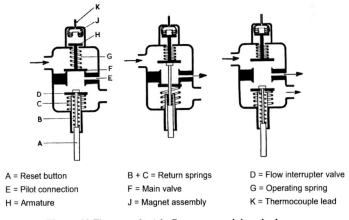

A = Reset button
E = Pilot connection
H = Armature
B + C = Return springs
F = Main valve
J = Magnet assembly
D = Flow interrupter valve
G = Operating spring
K = Thermocouple lead

Figure 13 Thermo-electric flame supervision device

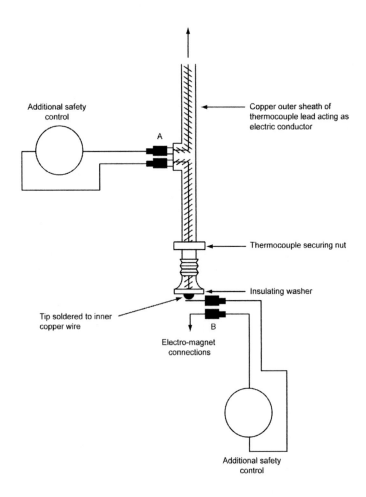

**Figure 14 Interruptible thermocouple
(connections made at either position A or B)**

Faults

- **Pilot flame**
  Partially blocked or incorrect position of pilot flame, which should play on the tip or the top 12 mm of the thermocouple. Under-aerated pilot flames are easily disturbed by draughts and have a low flame temperature.

- **Thermocouple**
  Contacts must be clean and tight (tight means about a quarter of a turn beyond hand tight). Short circuits can be caused by distortion of insulating washer by overtightening of the union nut. The tip should be clean and undamaged. The tip should be kept below red heat if the thermocouple is to have a reasonable life.

- **Electromagnet**
  Failure rarely occurs of the armature and magnet, which are housed within a sealed unit. Unit exchange is required when this happens.

Testing thermo-electric flame supervision devices (FSD)

There are numerous procedures for determining whether these devices have shut down correctly. The factors which determine which procedure to adopt are:

- the availability of pressure testing nipples on the equipment involved
- the accessibility of the burner or pilot burner being controlled (i.e. open or room-sealed flues)
- whether the FSD shut-off can be proved by including the procedure within a tightness test of the complete gas installation using the test nipple on the gas meter.

Where the manufacturer's instructions do not specify a procedure then the following may be used:

1. Ensure that all primary and secondary controls are set so that the burner will not be turned off during this procedure.
2. Light the appliance, and allow the burner to reach its normal working temperature.
3. Turn off the appliance shut-off device (appliance isolation valve) and simultaneously start a stopwatch.
4. Halt the stopwatch when the valve is heard to close.
5. Immediately check that the valve in the FSD has shut off completely, using the most appropriate method as indicated below:

   - **Preferred option**
     Where the appliance gas control system has a test nipple upstream of the FSD device (normally defined as test point P1), test for tightness and let-by between the appliance isolation valve and the FSD. If the FSD device has not shut off completely then a drop in pressure will occur.

- **Option 2**
  Where it is possible to complete a tightness test at the meter, or other suitable position upstream of the appliance isolation valve, then by turning on the appliance isolation valve the integrity of the FSD is also included. (Procedures must be adopted to ensure that any escape indicated by the gauge is not elsewhere on the gas installation pipework other than the FSD.)

- **Option 3**
  Where neither of the above options are available and the appliance is open flued, or the pilot and main burner are readily accessible, then where possible connect a gauge to the burner test nipple and turn on the appliance isolation valve. If any apparent increase in pressure is observed immediately turn off the isolation valve as this indicates that the FSD has not shut off completely. If no apparent increase in pressure is observed (or no gauge has been connected) immediately check with a lighted taper that gas has been interrupted to the main and pilot burner.

6. Check that the time recorded by the stopwatch conforms with the current requirements for gas appliances of heat inputs below 70 kW.

Oxygen Depletion System (ODS) – Atmosphere Sensing Device (ASD)

The European Gas Directive, 1 January 1996, states that when undergoing type testing to obtain the CE mark, appliances connected to a flue for the dispersal of combustion products must be so constructed that in abnormal draught conditions there is no release of combustion products in a dangerous quantity into the room concerned.

Domestic gas appliance design allows for excess air under normal operating conditions to be entrained into the appliance combustion chamber and hence to the atmosphere via the flue. When there is a spillage of combustion products into the room where the appliance is installed, complete combustion will occur for a period even though the oxygen level is decreasing and the carbon dioxide level is rising. However, as the oxygen level falls further, incomplete combustion occurs and carbon monoxide (CO) starts to be produced. The appliance design is such that the rate of CO production is initially low as the oxygen level falls and it is at this point that the oxygen depletion system (ODS) within the appliance intervenes.

A typical ODS (S.I.T. Gas Controls Limited) uses a controlled flame to heat a thermocouple, being part of a thermo-electric flame supervision device. As the oxygen level decreases in the atmosphere, so this controlled flame 'lifts' in search of oxygen, thus reducing the heat applied to the tip of the thermocouple until at a pre-determined point the electric current is reduced sufficiently to shut off the gas supply to the appliance (figures 15, 16 and 17).

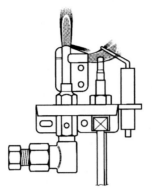

Figure 15 Adequate oxygen supply

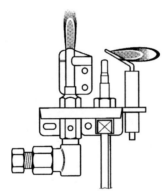

Figure 16 As the oxygen level falls, the sensing flame lifts away from the thermocouple tip

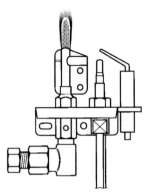

Figure 17 Just prior to shutdown – the sensing flame has completely extinguished

The ODS has an intervention level of 200 ppm (0.02%) of CO concentration in the room in which the appliance is installed.

The installation and annual servicing of all appliances must be conducted by competent operatives and those checks and tests to prevent incomplete combustion occurring as specified in Regulation 26(9) of the current Gas Safety (Installation and Use) Regulations must be complied with.

The ODS device must be checked according to the manufacturer's instructions whenever work has been carried out on the appliances in addition to any annual safety checks. The main points of these checks are to ensure:

- no part of the ODS is damaged

- the ODS is securely mounted in its recommended location

- the flame picture is not distorted and is burning correctly at the main burner cross-ignition port, the sensor port and its inter-connecting ribbon burner

- the aeration port adjacent to the injector is free from any obstruction.

If a customer reports that the ODS keeps 'going out' there is a high probability that it is working correctly and doing exactly what it is supposed to do by making the appliance safe in the event of progressive oxygen starvation due to abnormal flue or ventilation conditions.

Note: The S.I.T. Oxypilot ODS has no serviceable components and if required a complete unit exchange is necessary (figure 18).

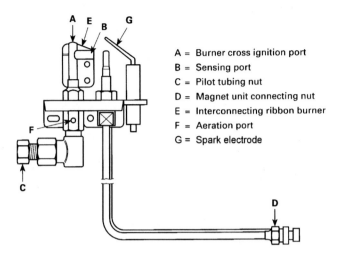

Figure 18 S.I.T. Oxypilot ODS

A = Burner cross ignition port
B = Sensing port
C = Pilot tubing nut
D = Magnet unit connecting nut
E = Interconnecting ribbon burner
F = Aeration port
G = Spark electrode

Flue Safety Devices

These devices are used to detect adverse flue conditions (spillage) at the draught diverter of an open flued appliance. They are known as TTB's (a Dutch acronym of the words 'Themische Tervgslag Beveiliging') but are often referred to as down draught thermostats, thermoswitches or smoke thermostats.

These devices (heat sensors) are located just inside the draught diverter, and are linked:

- in series with the thermocouple of a thermoelectric flame supervision device, or
- to shut off the main burner solenoid valve, or
- to the electronic circuit board.

The sensors are pre-set and calibrated to avoid nuisance shutdowns while still maintaining safe tolerances. They require manual intervention to re-establish the gas supply to the main burner.

Flame Rectification

This method of flame supervision superseded the more basic flame conductance system, which was prone to simulated d.c. flame signals. Condensation or a build up of carbon, due to flame chilling, can bridge the probe and burner. With a d.c. signal where electrons travel around the circuit in only one direction, a control unit can not distinguish the presence of a flame, from the bridging of the gap between the probe and the burner. However an a.c. signal can, due to the two directional flow to and from the control unit.

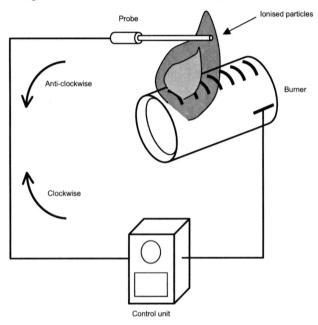

Figure 19 Flame rectification circuit

If we imagine the electrons from the control unit's signal, travelling in a clockwise direction, when the signal reaches the probe, the electrons are able to travel to the burner due to the ionised particles in the gas flame. If there was no flame present, the electrons are not supplied with sufficient pressure to jump the gap e.g. voltage/spark. The probe passing the electrons to the burner is very much like a shotgun firing pellets at a barn door (the burner has a much greater area). Therefore all of the electrons will travel the gap and be registered at the end of the clockwise journey back to the control unit.

When travelling back in the anti-clockwise direction, the electrons now try to pass the gap from the burner to the probe, this is now like shooting a cannon at a pencil, only some of the electrons are able to 'hit' the probe and travel back to the control unit. We now find a rectified signal recognised by the control unit as the presence of a flame.

The flame rectification system can distinguish various signals, for example:

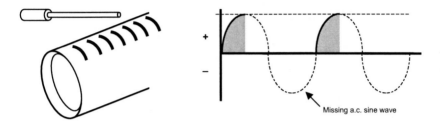

Figure 20 Open circuit

Figure 20, shows the signal read by the control unit where no flame or bridge is present. The electrons reach the end of the probe but have nowhere to go.

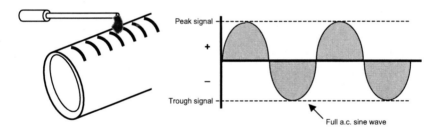

Figure 21 Closed circuit

Figure 21, shows the signal read by the control unit when the gap between the probe and burner, is bridged by conductive matter (condensation or carbon). The electrons can travel freely in both directions.

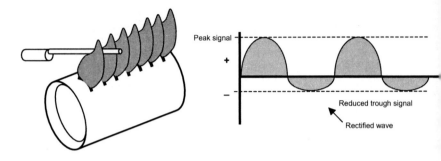

Figure 22 Rectified signal

Figure 22, shows the signal read by the control unit when the gap between the probe and burner, is bridged only by the flame, the shot gun and barn door effect now takes place, rectifying the signal.

A typical, burner head to probe ratio of 8:1 will allow a rectified signal to be produced. This ratio is easily surpassed for most atmospheric bar type burners.

Burner

These are normally stainless steel, multi-bar with preset aeration that cannot be adjusted. Each burner is usually fitted with its own injector.

Location of Components

The layout of the components of a single point instantaneous sink water heater are shown in Figure 23 and Figure 24. A multi-point heater with a thermostat is shown in Figure 25.

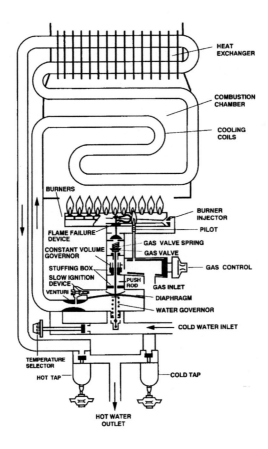

Figure 23 Instantaneous sink water heater

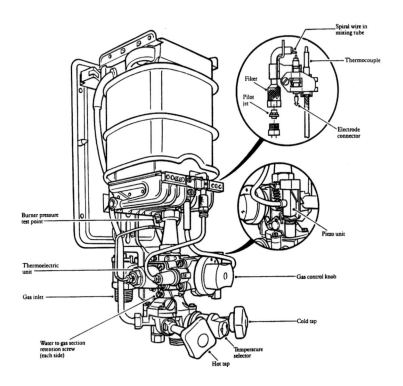

Figure 24 Component location and pilot burner detail

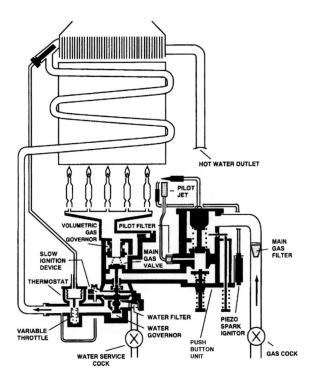

Figure 25 Multi-point water heater

INSTALLATION OF INSTANTANEOUS WATER HEATERS

Advice on materials, design and installation of water heaters is given in BS 5546: Specification for installation of gas hot water supplies for domestic purposes.

Further advice can be found in BS 6700: Specification for design, installation, testing and maintenance of services supplying water for domestic use within buildings and their curtilages.

Any work that is carried out on water installations in the United Kingdom must conform to the Water Supply Byelaws. These byelaws cover items relevant to the installation of hot water appliances such as:

- acceptable materials used for drinking water supplies
- protection against contaminated water entering the main water supply
- cold water storage protection
- identification of water pipes
- unvented hot water systems.

Before installing any gas appliance, the manufacturer's instructions should be consulted. They give specific directions for installation, including location requirements and minimum dimensions with regard to flue terminal positions.

For some appliances you may find the measurements given differ slightly from those stated in the British Standards. In this case the manufacturer has designed the appliance to operate satisfactorily with regard to specific installation instructions which should be adhered to.

The installation must conform to the current Gas Safety Regulations and, if the installation contains an electrical supply, to the Institute of Electrical Engineers Regulations (IEE).

The choice of a particular type of system, instantaneous, single point or multi-point, should match the requirements of the customer, taking into account acceptable positions for installation, availability of water and comparative installation and maintenance costs.

LOCATION GUIDELINES

Although the manufacturer's instructions give precise information regarding acceptable locations for appliances, the following points should be considered before selecting the exact location:

- suitability of the site
- route of any flue system
- acceptable terminal position
- available ventilation, if required
- pipework route for water supplies
- pipework route for gas supplies
- suitable electrical supply, if applicable
- access for maintenance and servicing
- protection against frost damage, if required.

SMALL SINGLE POINT INSTANTANEOUS WATER HEATERS

This type of appliance is normally flueless, except for instantaneous water heaters that are intended to be used continuously for periods in excess of 5 minutes.

To prevent staining of the ceiling above, the appliance should be fitted at a height that gives adequate clearance between the flue products outlet and ceiling.

The manufacturer's instructions should give specific measurements for the correct installation of the appliance, but where no guidance is given the following measurements can be referred to:

- minimum height above flueway to ceiling: 300 mm (12 in)
- minimum height above flueway to a combustible shelf: 150 mm (6 in).

If the water heater is to be connected to a flue system, there must be adequate space above the appliance to allow the fitting of the draught diverter and flue system. Detailed recommendations are given in BS 5440-1.

Sink water heaters should be positioned so that the outlet of the spout terminates not less than 13 mm (½ in) above the top edge of the sink or basin.

In circumstances where the heater provides hot water to a single draw-off via a pipework system, the length of pipework from the appliance to the draw-off should not exceed 3 m (10 ft). This will avoid undue cooling of the water before delivery.

INSTANTANEOUS MULTI-POINT WATER HEATERS

Room sealed appliances should be installed in an appropriate position for the location of the terminal.

Manufacturer's instructions must be consulted for specific installation guidance.

MAINS FED SYSTEMS

The water pressure and flow rate should be checked if a large instantaneous water heater is to be connected directly to the main. Combined water pressure and flow rate gauges are available.

Specific guidance on the minimum and maximum permitted flow rate and pressure can be found in the manufacturer's instructions.

CISTERN FED SYSTEMS

The minimum head of water available at the highest draw-off tap must conform to the manufacturer's instructions so as to allow the heater to operate with normal rates of flow of approximately 8 litres/minute (Figure 26).

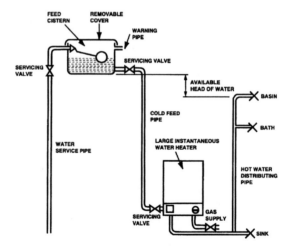

Figure 26 Cistern fed multi-point installation

FLUES AND VENTILATION

An adequate supply of air is essential for a gas appliance to burn safely and efficiently. For flueless and open flued appliances an air supply is also required to ensure adequate ventilation in the room or space in which the appliance is located.

The air requirement for a given appliance is dependent upon its rated heat input, whether it is flueless, open flued or room sealed, and its location in a room or a compartment. Room sealed water heaters are permissible in any room or internal space, provided that the flue terminal can be located as recommended by the manufacturer or BS 5440-1.

In new installations the following water heaters must be flued:

- large instantaneous (in excess of 12 kW input)
- small instantaneous heaters that are used for supplying showers, or other taps not located in the same room or space as the heater, or intended to be used for periods in excess of 5 minutes
- any water heater supplying a bath.

The following gives guidance on acceptable locations for a water heater with regard to flueing and ventilation.

More detailed information on flueing and ventilation of gas appliances can be found in the ConstructionSkills publication *Gas Safety*.

Bathrooms and Shower Rooms

Under no circumstances shall any open flued or flueless water heater be installed in a room containing a bath or shower.

Only room sealed appliances may be fitted in these locations, provided that the flue terminal can be located as recommended in BS 5440.

Where a room sealed water heater, with an electrical supply, is located in a room containing a fitted bath or shower, the mains electrical switch should be so sited that it cannot be touched by a person using the bath or shower, in accordance with the Institute of Electrical Engineers Regulations.

Living Rooms, Kitchens, Utility Rooms, Halls and Passageways

When selecting a living room as the location for a water heater, careful consideration should be given to the effect on the living area of such an installation, e.g. noise of the appliance, general aesthetics of the installation and servicing requirements.

Cloakrooms and Toilets

All types of water heater are permissible provided that the flueing and ventilation meet the requirements of BS 5440. Flueless water heaters should not be installed where the volume of the room is 5 m^3 or less. Any air vent must communicate directly with the outside air.

Compartments and Cupboards

A compartment is defined as an enclosed space within a building, either constructed or modified specifically to accommodate the water heater and its ancillary equipment.

The regulations do not permit flueless water heaters being installed in compartments. Room sealed water heaters may be fitted in this location, providing that the flue and ventilation requirement are as recommended by BS 5440-1 and -2.

Where a compartment is used to accommodate a flued water heater the compartment should conform to the following requirements:

- be of a rigid structure in accordance with any manufacturer's instructions regarding the internal surfaces

- combustible internal surfaces should be at least 75 mm (3 in) from any part of the heater or be suitably protected

- must have access to allow inspection, servicing and removal of the appliance and ancillary equipment

- the compartment must incorporate air vents for ventilation and, where necessary, for combustion as recommended in BS 5440-2

- no air vent must communicate with a bathroom, shower room, bedroom or bedsitting room, if an open flued appliance is fitted in the compartment.

Under Stairs Cupboards

This location should only be considered as a last resort in a building that is more than two storeys high. Wherever possible only room sealed appliances should be installed. Air vents must be direct to the outside air.

The cupboard should be treated as a compartment and, in addition, in a building that is more than two storeys high all the surfaces should be lined with non-combustible materials unless inherently fire resistant, e.g. plastered ceiling.

Bedrooms and Bedsitting Rooms

Instantaneous water heaters of more than 14 kW input must be room sealed.

A room with a volume exceeding 20 m^3 can be fitted with a small flueless water heater with a heat input not exceeding 12 kW (41,000 Btu/h) serving a wash basin or sink only.

Open flued water heaters of input not exceeding 14 kW can be fitted.

Both open flued and flueless water heaters must incorporate a safety control designed to shut down the appliance before there is a build-up of a dangerous quantity of the products of combustion in the room concerned.

Normal ventilation requirements for all water heaters must be complied with.

When installing a water heater in a bedroom or bedsitting room, consideration should be given to the factors affecting amenity, e.g. noise of operation.

Roof Space Installations

Where no other alternative exists and where local regulations permit, an open flued or room sealed appliance may be fitted, provided the appliance is protected from frost damage.

When this location is selected the roof space should have:

- flooring of sufficient strength and area to support the appliance and facilitate servicing
- enough vertical clearance when siting the cold water cistern to ensure the availability of the static head required by the appliance
- a suitable means of access to the heater, e.g. foldaway loft ladder, and sufficient fixed lighting
- a guard fitted to prevent any items stored within the roof space coming into contact with the appliance
- a guard-rail around the access hatch
- gas isolation control outside of the loft area.

WATER SUPPLY

Where the hot water system is supplied with water from the mains of a water authority, the design and installation must be in accordance with the Water Byelaws.

Materials

All pipes, fittings and materials used for the water supply must comply with the relevant British Standard and be listed in the UK Water Fittings and Materials Directory.

Material containing lead or other substances that constitute a toxic hazard shall not be used on water supplies intended for domestic use.

Byelaw 7 – Material

This byelaw prohibits any material likely to cause contamination of potable water for drinking purposes. This prevents the use of lead solders, hemp and linseed oil based jointing compounds. PTFE tape and lead-free solders should be used instead.

Byelaw 9 – Lead Pipe

Prohibits the installation of lead pipe.

Byelaw 10 – Copper and Lead

This byelaw prohibits the use of copper when repairing lead pipe to prevent corrosion through galvanic action.

Pipe Sizes

Water supply pipes to or from any water heating appliance should be at least the same size as the integral water connections on the appliance, or as recommended by the manufacturer.

In some instances, however, it may be necessary to calculate pipe sizes. Consideration should be given to the length of run, the number and type of fittings and the head of water available.

Full details of such calculations can be found in BS 6700.

Pipe Supports and Fixings

Pipes should be secured by purpose-designed pipe supports or clips at intervals not greater than those given in Table 1, and in such a manner as to allow free movement for expansion and contraction, particularly in long runs. Ensure that sufficient space is left between pipes if they are to be covered in an insulation material.

Table 1 Maximum interval between pipe supports

Nominal pipe size		Interval for vertical runs		Interval for horizontal runs	
(mm)	(inches)	(m)	(ft)	(m)	(ft)
Up to and including 15	Up to and including ½	1.8	6	1.2	4
22	¾	2.4	8	1.8	6
28	1	2.4	8	1.8	6
35	1¼	3.0	10	2.4	8
42	1½	3.0	10	2.4	8

Water Pipework – General Guidelines

The pipework should not be located where it is liable to be damaged, and should be fixed to avoid accumulations of dirt and to facilitate cleaning.

Where the pipework is fixed to the inside face of an external wall, it should not be in contact with the wall.

Hot water pipes must be free to expand and contract and therefore, wherever possible, they should not be buried in solid floors.

If pipework is laid in channels in floors and walls there must be adequate protection against damage by frost and corrosion, and no joints may be buried. Pipework covered in a protective coating is available, e.g. plastic coated copper pipe.

Where pipework is encased in a wall, a suitable duct with a removable cover should be provided.

Byelaw 58 – Accessibility

This byelaw states that where pipes are laid in floor screed, walls, or under suspended floors they must be reasonably accessible and, if necessary, laid in ducts.

Where pipes would be in contact with material likely to cause corrosion they should be protected by wrapping, or coated with a corrosion resistant material.

Pipework passing through a wall or ceiling should be sleeved with a suitable material (i.e. PVC) to allow for expansion and to facilitate replacement or repair.

Where it is necessary to run underfloor pipes at right angles to the joist, the depth of notch cuts in the joists should not exceed 15% of the joist depth. No notch should be further away than one quarter of the span from an end support.

Consideration should be given to protect the pipework, e.g. by covering the section passing across the joist with a protection plate to prevent floorboard fixings piercing the pipework.

Before the appliance is connected to the water supply any debris, i.e. flux and copper filings, should be flushed out of the pipework.

In order to reduce the quantity of water which is wasted before hot water flows from any draw-off, the length of water-distributing pipe should be kept as short as possible, and insulated. The maximum permissible length of uninsulated hot water pipe is given in Table 2.

Table 2 Maximum permissible length of uninsulated hot water pipes

Outside diameter mm (in)	Maximum length	
	m	ft
Up to and including 12 ($^3/_8$)	20	66
Over 12 ($^3/_8$) up to and including 22 ($^3/_4$)	12	40
Over 22 up to and including 28(1)	8	26
Over 28	3	10

Back Siphonage

To guard against the risk of back siphonage an air gap must exist between the outlet of the spout or tap and the maximum water level in the receiving vessel (Figure 27).

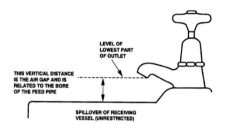

A 15 mm (½ in) draw-off tap should have a 20 mm air gap
A 22 mm (¾ in) draw-off tap should have a 25 mm air gap

Figure 27 Draw-off to bath or sink

Frost Protection

Storage vessels and pipework fitted in exposed places or roof spaces should be protected against frost and cold draughts by the use of insulating material.

Byelaw 49 – Protection from Freezing

This byelaw gives requirements for insulation and specific guidance on thickness and grades of insulating materials.

Special care is required with pipe positioning and insulation of pipes and tanks.

Scale and Corrosion

Hard water may deposit scale in pipes, fittings and the waterways in appliances, but does not normally cause corrosion.

Excessive deposits can cause restriction of water flow and overheating in appliances. The amount of scale deposited depends upon the temporary hardness present in the water and the temperature to which it is heated; scale can be minimised by design features which limit the water temperature, or by the use of proprietary chemical treatment methods. Where a circulator/storage heater is fitted in a hard water area it should only be connected to an indirect cylinder.

Naturally soft, or chemically softened water, can cause severe chemical corrosion to unprotected iron and steel and, in areas where such water is supplied, all water-carrying components should be of a non-ferrous material or be suitably protected.

Electrolytic corrosion tends to occur wherever certain dissimilar metals are in contact with water, the severity of the corrosion depending on the nature of the water and the temperature conditions.

In general, except for cold storage cisterns and feed cisterns, galvanised steel and copper should not be used together in the same system. Another cause of corrosion in galvanised water storage vessels is the presence of debris lying on the bottom. This may consist of drilling swarf, builder's debris, etc., and such debris should be removed.

Scale Treatment

Instantaneous water heaters installed in hard water areas can suffer from the problem of scale. The scale is formed from the build-up of layers of calcium salt, and the rate at which it is formed increases with the rise in water temperature.

Descaling of heating bodies may have to take place every two years in areas with a particular hard water problem. A method of descaling is shown below.

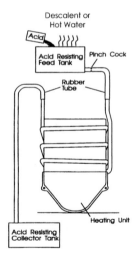

Figure 28 A method of descaling

When descaling the removed heating body is inverted and placed on a flat surface. It is then connected to an assembly of rubber tubes and acid resisting tanks as shown.

The upper tank will contain either a proprietary brand of descalent or a solution of hot water and hydrochloric acid. This fluid should be allowed to pass slowly through the heating body into the collecting tank. The process is continued until the solution ceases to bubble. The heating body is flushed with water thoroughly to clean out any residual solution, and may then be refitted to the water heater.

GAS SUPPLY

Copper tubing in accordance with EN 1057 is preferred for the gas supply, but steel pipe in accordance with BS EN 10255 may also be used.

The appliance must have an adequate gas supply and the pipework for this should be at least the same size as the appliance connection, or as recommended by the manufacturer.

In cases where a complete gas supply pipe has to be installed, the size of pipe selected must be of sufficient diameter to supply all the appliances that are on the installation at their maximum gas rate.

Consideration must be given to the pressure loss in each section of the pipework system and it is recommended practice that the pressure loss between the meter outlet and each appliance connection does not exceed 1 mbar.

A detailed method of calculating the correct pipe size for an installation can be found in the ConstructionSkills publication *Gas Safety*.

Protection and Jointing

Any pipework that is laid in a solid floor or wall, or passes through a fireplace opening, or in any location where corrosion may take place, must be protected to a distance of 25 mm (1 in) beyond the surface likely to cause the corrosion.

This may involve either:

- using pipework with a bonded coating of PVC
- using a protective sleeve or duct
- wrapping the pipework with a corrosion resistant tape.

Before any protective coating is applied the installation should be tested for tightness.

Compression fittings and union joints should only be used for a gas supply where they can be accessed readily to ensure a tight joint. They should not be used under floors or in ducts.

Unsoldered capillary joints have been known to satisfy the gas tightness test but begin leaking when the flux eventually melts. Finished joints should, therefore, be visually examined to ensure that the joint has been made.

When using a blowlamp on a gas pipe that contains gas, or has contained gas, the gas supply to that part of the system must be isolated and disconnected to prevent any possible ignition of the gas. Care should be taken that any combustible materials in the vicinity of the work area are removed and the area has been properly ventilated.

Temporary Continuity Bonds

When breaking into a gas supply, or disconnecting a supply, it is possible to isolate part of the supply from the electrical equipotential bonded zone. This could create a spark or inflict an electrical shock. There have been reported cases where a person has cut into pipework and an electrical current has passed from the pipe through their body to earth. Equipment is available, called a temporary continuity bond, which will bridge the section of pipe being removed and allow any current to pass through the pipework, but not through the operative.

The use of a temporary continuity bond is a requirement of the current Gas Safety Regulations.

Further information on general installation practices for internal gas supplies can be found in the ConstructionSkills publication *Gas Safety*.

ELECTRICAL SUPPLY (WHERE APPLICABLE)

All wiring should comply with the Institute of Electrical Engineers Regulations for electrical installations.

Electrical connections to the water heater and any ancillary controls should be carried out in accordance with the manufacturer's instructions.

All electrical components and wiring should be suitable for the maximum current rating. The connection to an existing mains electrical supply should provide means to isolate the appliance and should utilise either a 13 amp 3-pin socket, preferably unswitched, or double-pole, switched and fused spur box. The fuse rating in the plug or fused spur must not exceed 3 amps, unless a higher rating is specified by the manufacturer.

The socket outlet or spur should be readily accessible and preferably be adjacent to the appliance (except where the appliance is located within a bathroom). It should supply the appliance only and be easily identifiable as so doing.

Any switch or electrical control in a bathroom, except of the cord pull type, must be situated so that it is normally inaccessible to anybody using the bath or shower.

This does not apply to equipment which is designed for use in bathrooms, e.g. power showers. Only purpose-built socket outlets, e.g. shaver points, shall be installed in a bathroom and there shall be no means of connecting portable equipment within the bathroom.

Electrical work is specifically controlled by the Electricity at Work Regulations 1989 and this requires installers to be competent in the electrical work they are undertaking.

Following the guidance of the Institute of Electrical Engineers (IEE) Recommendations ensures compliance with the Electricity at Work Act. Installers should only undertake electrical work for which they have been suitably trained. Any installer who is not fully conversant with electrical work is recommended to call on the services of an electrician approved by The National Inspection Council for Electrical Installation Contracting (NICEIC).

All new metallic pipework, gas or water, installed in a bath or shower room, that can be touched at the same time as any other metallic fixture in the bathroom, must be provided with supplementary bonding in accordance with IEE Regulations.

This will require the new pipework to be bonded together and to an existing pipe within the room provided that pipe has been previously bonded correctly. Earth clamps and earth conductors of a minimum cross sectional area of 4 mm^2 may be used for this purpose.

INSTALLATION OF SHOWER UNITS

There are now many types of shower on the market that are suitable for connection to instantaneous water heaters.

GRAVITY SHOWERS

Used in conjunction with a hot water storage cylinder and where both the cold water and hot water are fed by gravity from a feed cistern (Figure 29).

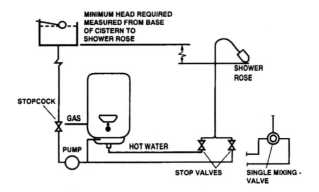

Figure 29 Shower installation, cistern water supply

MAINS PRESSURE SHOWERS

The hot water is supplied to a mixing valve from either a mains fed instantaneous water heater, combination boiler or unvented hot water storage system. The cold water connection to the mixing valve is also taken directly from the mains water supply (Figure 30).

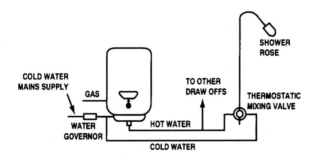

Figure 30 Shower installation, mains water supply

PUMPED SHOWERS

This group of shower systems can be split into those having the pump integral with the shower control unit and mixing valve, and those where the pump is located in a position remote from the control unit.

The manufacturer's installation instructions should be adhered to for specific models, and in general the installation should comply with all relevant regulations regarding gas, water and electrical installations as previously mentioned.

In addition, BS 6340 should be consulted.

SHOWER UNITS – GENERAL GUIDELINES

A flexible shower hose must have a restraining mechanism that prevents the handset discharging any closer than 25 mm above the lowest part of the top edge of the bath or shower tray, unless a double check valve is fitted which meets the requirements of the water supply (water fittings) regulations.

Pipe runs to showers should be kept as short as possible. To reduce the possibility of temperature variation and risk of scalding both hot and cold supplies should come from the same source.

Where multi-point water heaters are connected direct to a mains water supply it is essential there is adequate working pressure and flow available.

In a cistern fed system, a check should be made to ensure that the cold water supply to the cistern provides a sufficient flow rate to keep the cistern topped up under normal use.

In most cases the hot and cold supply pipes to a gravity-operated system should be 22 mm with the exception of the final connection to the mixer. Pipework to pumped or mains pressure showers should normally be 15 mm.

Hot and cold supplies should be installed to ensure that the flow rate to the shower does not appreciably reduce when other outlets are used.

It is preferable to connect direct to the cold water cistern for a cold water supply to a shower. Hot water supplies should be fed as an independent supply from the cylinder, e.g. by utilising an Essex or Surrey Flange.

When sink heaters are used to operate showers the following points should be noted:

- the heater must be flued
- it should not be installed in the same room as the shower
- it should only supply one shower.

INSTALLATION OF WASHING MACHINES AND DISHWASHERS (HOT FILL)

Instantaneous multi-point water heaters may be connected to suitable automatic washing machines (Figure 31), but single point sink water heaters are considered unsuitable, because of the low water flow rate.

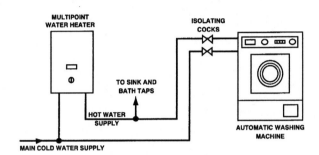

Figure 31 Connection to a washing machine

Instantaneous multi-point water heaters may also be used to provide hot water to dishwashers, but the manufacturer's instructions should be consulted for advice on compatibility and specific installation requirements.

The washing machine/dishwasher and water heater should be connected to the same mains water supply.

Some dishwashers are now designed to operate on a 'hot fill' basis for use with water supplied from a hot water storage system.

The washing machine or dishwasher should conform to the following requirements:

- be installed to comply with the Water Supply (Water Fittings) Regulations
- be controlled by a pressure level switch which cuts the water supply when the correct level has been reached
- should not operate on a 'timed fill cycle'
- should not use a mixing thermostat where the hot and cold solenoids are rapidly switched on and off to mix the water
- should contain an electrical heating element to boost the water temperature for programmes requiring water above 60°C.

COMMISSIONING CHECKLIST FOR INSTANTANEOUS WATER HEATERS

It should be noted that the checklist shown may be used in the absence of the appliance manufacturer's instructions.

On completion of the installation, both inlet gas and water supplies should be tested for tightness and purged, and a flue flow test applied to any open flued system. The commissioning procedure below may then be followed.

1. Ensure installation meets all relevant regulations.
2. Check ventilation is adequate, where appropriate.
3. Check for water leaks.
4. Check ignition device and light pilot.
5. Check pilot flame length and position, and flame supervision device operation.
6. Check for gas tightness on heater joints.
7. Ensure heater lights when the hot tap is turned on.
8. Check burner pressure corresponds to data badge.
9. Check the gas rate, if necessary, and the flame picture.
10. Check temperature rise and water flow rate and adjust where necessary.
11. Check the thermostat, if fitted.
12. Check the draught diverter on an open flued sink water heater for spillage.
13. Check points and seals on room sealed appliances.
14. Check slow ignition device is adjusted correctly by turning the tap on and off.
15. Ensure that a multi-point water heater operates on all taps.
16. Ensure that any appliance warning label is fitted, where appropriate.
17. Instruct the customer in the use of the appliance and on the need for regular servicing.

MAINTENANCE CHECKLIST FOR INSTANTANEOUS WATER HEATERS

It should be noted that the checklist shown may be used in the absence of the appliance manufacturer's instructions.

PRE-SERVICE CHECKS

1. Check with the customer that the water heater is working satisfactorily.
2. Check that the appliance is correctly fitted, in a suitable location, and that the room size is adequate.
3. Check that there is adequate ventilation with a satisfactory flue route and termination point for open flued appliances.
4. Check there are no signs of spillage on the appliance or adjacent walls.
5. Check the operation of all controls, the flame supervision device and ignition system.
6. Check flame picture.
7. Check the flow rate and temperature of water.
8. Check for any signs of water leakage.
9. Advise the customer of any problems.

FULL SERVICE

1. Isolate gas, water and, where necessary, electricity supply.
2. Remove outer case, clean out deposits and ensure all case seals are intact.
3. Drain heater by opening hot taps and by removing the drain plug.
4. Disconnect and remove heating body, and examine its condition. Clean the flueways and carry out a flue flow test on open flued systems.
5. Remove and clean main burner and injectors.
6. Remove and clean pilot burner and injector.
7. Examine condition of electrodes and thermocouples.
8. Where applicable, remove and clean the fan and check any air pressure sensing tubes (fan-flued appliance).

9. Ensure that the gas valve spindle moves freely and there is no sign of water leakage.

10. Check water and gas filters.

11. Ease and grease gas taps if necessary.

12. If removal of the water section is necessary for the replacement of the diaphragm on a multi-point, a temporary continuity bond should be fitted across the water supply.

13. Replace all washers, gaskets and case seals as required.

14. Reassemble, restore and check gas and water for tightness.

15. Check ignition device and light pilot.

16. Check pilot flame and test flame supervision device.

17. Ensure that the automatic valve operates correctly and, if necessary, adjust the slow ignition device.

18. Ensure the working pressure, gas rate and flame picture are correct.

19. Replace the outer case.

20. Check the draught diverter on an open flued sink water heater for spillage.

21. Check temperature and flow rate at the draw-off.

22. Ensure appropriate warning labels are attached.

23. Leave the appliance in working order and advise the customer of any further work required.

FAULT-FINDING CHECKLIST FOR INSTANTANEOUS WATER HEATERS

Symptom	Possible cause	Action
1. Pilot will not light	a) Gas not turned on	a) Turn on the gas at the gas service cock and/or the main gas line
	b) Air in gas line	b) Purge the gas line by depressing and holding the centre ON button, this may take 2–3 minutes
	c) Incorrect pilot lighting procedure	c) Follow the lighting instructions or refer to the 'Instructions for Use'
	d) Electrode lead not connected to the rear of the spark igniter	d) Reconnect the electrode lead
	e) Incorrect spark gap	e) Refer to manufacturer's instructions for correct spark gap and adjust
	f) Current tracking to earth	f) Check the electrode lead is routed clear of all metal parts
	g) Pilot injector blocked	g) Clear injector by blowing through or replace with new pilot injector. Clean the lint filter if fitted
	h) Faulty spark generator or electrode assembly	h) Replace faulty components
2. Pilot lights but will not remain alight	a) Thermocouple connections loose	a) Tighten the connections
	b) Pilot flame unstable or too small to heat the thermocouple tip	b) Pilot injector partially blocked, clear by blowing through or replace. Check the gas inlet pressure is correct – 20 mbar. Clean the lint filter if fitted
	c) Thermocouple worn out or damaged	c) Replace the thermocouple

Symptom	Possible cause	Action
	d) Faulty magnet unit in the flame safety device	d) Replace flame safety device
	e) Excessive draughts due to faulty seals	e) Replace seals
3. Pilot established but main burner will not light on water flow	a) Gas inlet pressure low	a) Check gas filter clear. Check pressure at meter, and if incorrect contact the gas supplier
	b) Low water flow caused by blocked water filter	b) Clean the debris from the filter
	c) Low water flow caused by scaled heat exchanger	c) Descale or replace heat exchanger
	d) Faulty diaphragm. Will also cause a high water rate	d) Replace the diaphragm assembly
	e) Automatic gas valve push rod jammed. May also cause a high water rate	e) Dismantle, clean and regrease. Handle the push rod with care, do not bend
	f) Slow ignition screw incorrectly set. Will also cause a high water rate	f) Adjust the slow ignition screw until the burner ignites smoothly and quietly when a hot water tap is turned on
4. Zero or low water flow rate	a) Blocked water filter	a) Clean the debris from the filter
	b) Heat exchanger blocked with scale	b) Descale or replace
	c) Loss of service water main pressure	c) Contact your local Water Authority
5. High water temperature	a) Automatic gas valve push rod sticking. Water rate will be normal	a) Dismantle, clean and regrease. Handle the push rod with care, do not bend

Symptom	Possible cause	Action
6. Low water temperature	a) Gas pressure too low	a) Check and clean gas filter, also check gas inlet pressure
	b) Faulty diaphragm. Will also cause a high water rate	b) Replace the diaphragm assembly
	c) Automatic gas valve push rod sticking. May also cause a high water rate	c) Dismantle, clean and regrease. Handle the push rod with care, do not bend
	d) Slow ignition screw incorrectly set. Will also cause a high water rate	d) Adjust the slow ignition screw until the burner ignites smoothly and quietly when a hot water tap is turned on
7. Noisy heater	a) Heat exchanger scaled. Ultimately this will cause the heat exchanger fins to discolour and buckle	a) Descale or replace heat exchanger
	b) Noisy ignition could be caused by incorrect setting of the slow ignition screw	b) Adjust the slow ignition screw until the burner ignites smoothly and quietly when a hot water tap is turned on
	c) Reduced pilot rate caused by dirt	c) Clear injector by blowing through or replace with new pilot injector. Clean the lint filter if fitted
	d) Burner aeration ports and main flame ports blocked	d) Carefully clean the burners with a vacuum cleaner
8. Smell of combustion products	a) Faulty case or terminal seal	a) Check that the outer case seal and terminal seal are in good condition, replace if necessary. Ensure the outer case is correctly positioned
	b) Failure to follow instructions with regards to openable windows and doors	b) Resite the appliance

Meters

CONTENTS

	Page
INTRODUCTION	1
DEFINITIONS	2
SIZING OF METERS	5
Diversity Factor	5
Determining Meter Sizes	5
METER LOCATION	8
General Requirements	8
Location Outside the Building	9
Location Inside the Building	12
Multiple Meter Installations	12
Mobile Dwellings (2nd Family Gases)	13
Meter Installations for Caravan Holiday Homes and Residential Park Homes	13
Meter Installations for Permanently Moored Boats	13
INSTALLATION OF METERS AND ASSOCIATED CONTROLS	14
Warning Notices	17
TESTING AND PURGING	20
Tightness Testing for 2nd Family Gases	20
Testing Procedures	20
Supporting Information for Testing where the Supply MOP Exceeds 75 mbar, But a Meter Inlet Valve (MIV) is Not Fitted	20
Soundness Testing for 3rd Family Gases	21
Results of Tightness/Soundness Test	21
Pass	21
Fail	21
Purging	22
Meter and Regulator Operation	23
REMOVAL AND EXCHANGE OF METERS	25

INTRODUCTION

This reference manual specifies the requirement for the safe installation of gas meters up to 6 m^3/h (212 ft^3/h) rating in domestic, industrial and commercial premises for gas at inlet pressures to the meter not exceeding 75 mbar. The installations supplied through low or medium pressure distribution systems of 2nd and 3rd family gases must not exceed a pressure of 2 bar.

The provision for service pipes and regulators are dealt with in the Institute of Gas Engineers Recommendations IGE/TD/4 Gas Services and the LPGA Codes of Practice No. 25 – LPG Central Storage and Distribution Systems for Multiple Consumers (March 1999).

The installation of pipework is dealt with in BS 6891 for 2nd family gases and BS 5482-1 and 2 and the LPGA Codes of Practice No 22 – LPG Piping System Design and Installation (February 1996) for 3rd family gases.

All meters must be installed in accordance with the current Gas Safety (Installation and Use) Regulations, and BS 6400: 2006 – Specification for installation, exchange, relocation and removal of gas meter with a maximum capacity not exceeding 6 m^3/h. Part 1 low pressure (2nd family gases), Part 2 medium pressure (2nd family gases) and Part 3 low and medium pressure (3rd family gases).

The Office of Gas Supply (OFGAS) have produced a Code of Practice for low pressure diaphragm and electronic meter installations with badged meter capacities not exceeding 6 m^3/h (212 ft^3/h).

For gas meters of rating in excess of 6 m^3/h (212 ft^3/h) reference should be made to the Institute of Gas Engineers Recommendations IGE/GM/6: 1996 Specification for low pressure diaphragm and rotary displacement meter installations with badged meter capacities exceeding 6 m^3/h (212 ft^3/h) but not exceeding 1,076 m^3/h (3,800 ft^3/h), and IGE/GM/1: 1998 Gas meter installations for pressures not exceeding 100 bar.

DEFINITIONS

Credit meter	A meter in which the volume registered by the index is the basis of a periodic account rendered to the consumer
Cross bond	An electrical conductor connected between a point close to the outlet of a meter and the earth terminal
Diaphragm meter	A positive displacement meter in which the measuring chambers have deformable walls, e.g. a U6 gas meter
Electrical insulator	A fitting having high electrical resistance inserted in the service pipe or at the meter outlet to minimise the flow of electrical current
Electronic (ultrasonic) meter	A meter that measures indirectly the volume of gas passing through it, e.g. an E6 electronic gas meter
Emergency control valve (ECV)	A valve intended for use by a consumer of gas for shutting off the supply of gas in an emergency
Gas meter	An instrument for measuring and recording the volume of gas that passes through it, without interrupting the flow of gas
Index	A series of dials or rows of figures indicating the volume of gas that has passed through the meter
Installation pipework	Any pipework or fitting that connects the meter outlet union connection to points at which appliances are to be connected
Limited capacity relief valve	A device, actuated by excess outlet pressure of the regulator, designed to permit a maximum of 5% of the rated capacity to be discharged to atmosphere at an outlet pressure not exceeding the lower limit of the OPSO and to reseat as the outlet pressure decreases, at a pressure in excess of the maximum lock-up pressure
Lock-up	The action of the regulator valve to seal and prevent an excessive rise in outlet pressure under conditions of zero flow
Low pressure	Gas inlet pressure to the meter regulator of not more than 75 mbar

Medium pressure	Gas inlet pressure to the meter regulator between 75 mbar and 2 bar
Meter box	A receptacle or compartment designed and constructed to contain a meter with its associated gas fittings
Meter bracket	A purpose made support incorporating a means of securing the meter unions from which a meter can be suspended
Meter compound	An area or room designed and constructed to contain one or more meters with their associated gas fittings
Meter inlet valve (MIV)	The valve fitted upstream of, and adjacent to, a meter to shut off the supply of gas to it. (In most cases, the meter inlet valve may act as the emergency control)
Meter filter	A filter fitted between the meter inlet valve and the meter. (These may be integral with the regulator unit to form a filter/regulator unit)
Meter housing	A meter box or meter compound
Meter rating	The maximum volumetric rate of flow at which the performance of a meter is certified to be within statutory requirements or some other agreed requirement. (This rate is indicated on the index of the meter)
Meter regulator (governor)	A device which controls the pressure at its outlet within predetermined parameters and which is normally fitted adjacent to and upstream of a meter
Non-return valve	A device to prevent the reverse flow of gas, or air or other extraneous gas
Overpressure shut-off (OPSO)	A manually resettable device that closes to prevent the flow of gas when pressure on the downstream side of the regulating member rises to a predetermined value. This device is an integral part of the regulator
Prepayment meter	A meter fitted with a mechanism which, on the insertion of a coin or token (mechanical or electronic) permits the passage of a predetermined quantity of gas

Pressure test point	A fitting provided for temporary connection of a pressure gauge
Primary meter	The meter nearest to and downstream of a service pipe or service pipework for ascertaining the quantity of gas supplied through that pipe or pipework by a supplier
Secondary meter	A meter, other than a primary meter, for ascertaining the quantity of gas provided by a person for use by another person, whether or not there is also a primary meter in respect of the gas supplied. For billing purposes these are sometimes referred to as sub-deduct meters
Semi-rigid stainless steel connector	A stainless steel tube formed with annular corrugations and having factory-fitted end connections
Service pipe	A pipe connected to a natural gas distribution main to provide a supply of gas to one or more customers and terminating at, and including the meter inlet valve at, a primary meter
Service pipework	A pipe connected to the first stage regulator of an LPG bulk tank to provide a supply of gas to one or more customers and terminating at medium pressure at a combined OPSO/UPSO, 2nd stage regulator and limited capacity relief valve, and the meter inlet valve

Service valve	A valve (other than an emergency control) for controlling a supply of gas, being a valve:
	    a) incorporated in a service pipe; and
	    b) intended for use by a supplier or transporter of gas; and
	    c) not situated inside a building

Thermal cut-off	A safety device designed to stop the flow of gas when the air temperature in the vicinity of the device has exceeded a predetermined value
Underpressure shut-off (UPSO)	A manually resettable device that closes to prevent the flow of gas when the pressure on the downstream side of the regulating member falls to a predetermined value. This device is an integral part of the regulator

SIZING OF METERS

DIVERSITY FACTOR

A diversity factor is given to each type of appliance according to the normal degree of intermittence of its use.

Table 1 Diversity factors of appliances for meter sizing

Type of appliance	Diversity factor
Central heating appliances (other than combination boilers)	1.0
Circulators	1.0
Refrigerators	1.0
Unit heaters	1.0
Combination boilers	0.8
Instantaneous multi-point water heaters	0.8
Sink water heaters	0.6
Hotplates	0.6
Ovens	0.6
Room heaters	0.6
Tumble dryers	0.6
Wash boilers	0.6
Cookers	0.4

DETERMINING METER SIZES

To calculate the size of meter required, multiply the maximum input rating (kW) of each appliance by its diversity factor. Multiply the sum of these heat inputs by 3.6 to convert to megajoules per hour, and divide by the calorific value of the fuel to obtain m^3/h.

Calorific value of natural gas is taken as 39 MJ/m^3

Calorific value of propane is taken as 93 MJ/m^3

so m^3/h = $\dfrac{\text{Total kW load} \times 3.6}{\text{CV of gas}}$

where total kW load = (appliance 1 x DF) + (appliance 2 x DF) etc.,

and DF = Diversity Factor

and CV = Calorific Value of fuel

Example:

Type of appliance	Heat input (kW)		DF		Load (kW)
Combination boiler	30	x	0.8	=	24
Cooker	21	x	0.4	=	8.4
Gas fire	7	x	0.6	=	4.2
Tumble dryer	3	x	0.6	=	1.8
Total load					38.4

m^3/h = $\dfrac{38.4 \times 3.6}{39}$

= 3.4 natural gas (2nd family)

m^3/h = $\dfrac{38.4 \times 3.6}{93}$

= 1.49 propane (3rd family)

Meter size

2nd family gas   =   U6 / E6 / G4

3rd family gas   =   U2.5

Typical domestic meters are indicated in Figure 1.

U6 meter

Coin or mechanical token meter

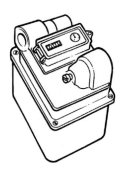
Semi-concealed meter

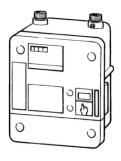
Electronic token meter

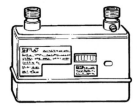

E6 meters

Figure 1

METER LOCATION

GENERAL REQUIREMENTS

Meters or associated controls shall not be located where:

a) food is stored

b) they might cause an obstruction

c) they might be exposed to accidental damage

d) they are in close proximity to any source of heat or may be subjected to extremes of temperature

e) they might be affected by damp or corrosive atmospheres

f) they will constitute a danger to any person

g) they will be in close proximity to electrical wiring, switchgear etc.

Prepayment meters shall be inaccessible to unauthorised persons, but in a convenient position for easy operation of the coin or token mechanism and withdrawal of any cash box.

Regulators and relief valves for medium pressure installations shall not be sited inside the building.

Meter installations for 3rd family gases shall not be located in basements or below ground level.

The meter shall be located so as to permit the installation, adjustment, servicing and repair of associated controls and the exchange of the meter itself, and shall be accessible for inspection and meter reading.

The meter and its associated controls shall not be in contact with any wall and shall be protected either by design, installation or position, from contact with any cement and/or cement composition and any floor that may be wetted.

Where gas and electricity meters and their associated controls are fitted within 150 mm of one another, a non-combustible partition made of an electrically insulating material shall be placed between them.

The following locations are suitable for gas meter installations:

- in a purpose made meter box or compound outside the building

- in a garage or suitable outbuilding

- inside the building

- at the boundary of the property, in a suitable enclosure.

LOCATION OUTSIDE THE BUILDING

External meter installations shall be located in meter housings that give adequate protection against the weather and acts of vandalism.

The size of the meter housing will be determined by the meter to be fitted and by the arrangement of the pipework and associated controls.

The meter shall be adequately ventilated. (Above ground housings should have a total effective vent area of 2% of the plan area of the housing divided equally between high and low level. Housings partly below ground level, e.g. semi concealed, shall have an effective vent area of 6% of the plan area at high level. All ventilation to be purpose designed.)

Only meters that are suitable for use with semi-concealed boxes shall be installed in such boxes. Semi-concealed boxes are not suitable for LPG meter installations.

Meter housing shall be so designed, constructed and installed that:

- customer access is gained only by a special key, and
- in the event of a gas leak inside the box, the gas cannot enter the building and/or cavity of its external wall.

Typical meter boxes are illustrated below. These are intended to accommodate the U6 and E6 credit type meters.

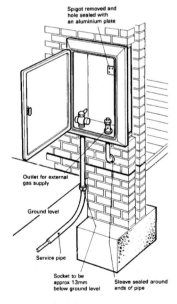

Figure 2a Gas service termination at approved built-in meter box

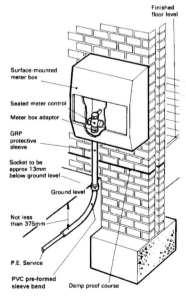

Figure 2b Gas service termination at approved surface-mounted box

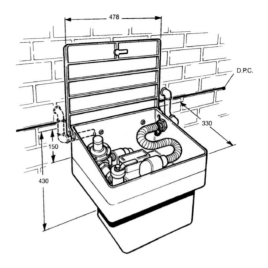

Figure 3 Gas service termination at approved semi-concealed meter box

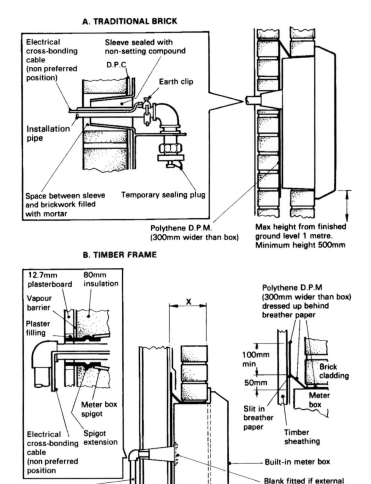

Figure 4 Installation of approved built-in meter box

LOCATION INSIDE THE BUILDING

The primary meter shall be located as near as practicable to the point where the service pipe or, if a multiple service is installed, where the service pipe branch enters the building or a flat in a block of flats.

In buildings of two or more floors above ground level no meter shall be installed on or under a stairway, in a common hallway, passageway or other position which provides the only means of escape in case of a fire.

In buildings of less than two floors above ground level it is preferable not to install any new or replacement meter on or under a stairway, or in any other part of the premises where the stairway or that other part of the premises forms the sole means of escape in case of a fire.

If this is unavoidable, the meter and its associated controls should:

- be of fire-resistant steel construction, or

- be housed in a fire-resistant compartment having automatic self-closing doors, or

- include a thermal designed cut-off that shuts off the gas immediately upstream of the meter, or regulator if fitted, if the temperature of the device exceeds 95°C.

Automatically operated light switches shall not be fitted to doors of cupboards or compartments where meters are located.

MULTIPLE METER INSTALLATIONS

Where a number of primary meters are grouped together that serve flats in large premises and flats over business premises, then the following provisions shall apply:

- each meter shall be clearly marked to indicate the flat or premises that it serves

- there must be reasonable access at all times to the area where the meters are located

- the meters shall be enclosed in a single, ventilated lockable compartment or individual meter boxes

- the occupant of each of the premises the meter serves shall hold a key for the meter housing.

MOBILE DWELLINGS (2ND FAMILY GASES)

Meter Installations for Caravan Holiday Homes and Residential Park Homes

Metering of gas to these situations may be either:

a) through a single site meter measuring gas to more than one mobile dwelling

b) through a primary site meter with individual secondary meters to each mobile dwelling. In these situations, where the meter regulator is set to provide the pressure so that 21 mbar (see also pressure at the meter outlet, page 23) is obtained at the inlet to each individual dwelling installation pipework, no regulator is needed at the secondary meter installation. Where the pressure is set above 21 mbar working pressure at the inlet to the secondary meters, then the secondary meters shall also have regulators fitted

c) individual primary meters to each mobile dwelling.

As near as practical to any site meter shall be displayed, in a readily accessible position as near as possible to the primary meter, a line diagram in permanent form. The diagram must show the location of all installation pipework of 25 mm or greater diameter, meters, emergency controls, valves and pressure test points.

Any individual dwelling meter shall be installed external to the dwelling in a securely fixed meter box or similar robust housing which allows adequate ventilation.

Meter Installations for Permanently Moored Boats

The gas supply to a permanently moored boat shall be metered as for mobile dwellings. However, any primary meter used shall be mounted on the bank above normal flood level. Where an individual meter is used as a secondary meter on the boat, then it shall be possible to isolate the vessel individually from a position which is unaffected by flood conditions and is away from the boat, for example on a separate floating pontoon or on the bank. Reference should be made to the appropriate Navigation/Water Authority.

INSTALLATION OF METERS AND ASSOCIATED CONTROLS

Diaphragm meters shall be kept upright at all times. When handling and transporting any meter, care must be exercised to avoid mechanical shock or damage.

Extreme care must be taken when handling electronic token meters. These meters have tamper devices which will prevent gas passing if unauthorised work is performed on a meter. If activated, the device will prevent a gas tightness test from being undertaken.

When a meter is not connected, the inlet and outlet connections shall be capped or blanked off.

The means of support and connection of the primary meter shall be acceptable to the gas supplier or transporter, which currently is by suspending the U6, G4 or E6 meter from a meter bracket, which provides a secure fixing to restrict movement and reduce tampering.

Methods of connection include:

- semi-rigid stainless steel
- mild steel
- copper.

The bending radius of the semi-rigid stainless steel connector, if used, should be limited so that corrugations are not close enough to allow condensation to bridge the gap which could result in corrosion. (A meter of 6 m^3/h capacity will meet the needs of the majority of 2nd family domestic gas installations.)

It should be noted that only one semi-rigid connector may be used on each meter installation. Unless the meter is securely restrained, e.g. by a meter bracket, the connection to the meter installation shall be in securely fixed rigid pipe for at least the first 600 mm.

The meter shall be either a credit or prepayment type. A prepayment meter shall not supply a secondary meter. A secondary meter shall carry a label stating 'secondary meter', and a label shall be fitted to its primary meter indicating the number and location of any secondary meters supplied through it.

The emergency control valve (ECV) shall be fitted to the service pipe as close as practical to the inlet of the meter regulator of the primary meter.

Where the maximum operating pressure (MOP) exceeds 75 mbar an additional valve known as a meter inlet valve (MIV) shall be fitted between the regulator and the meter.

An additional emergency control valve (AECV) shall be fitted as close as practicable to every secondary meter.

The emergency control valve shall be sited so as to permit easy access for operating, servicing and exchanging and have a key or lever so fitted that the lever cannot be moved in a downward direction to the 'open' position. Any detachable lever shall be securely held in place.

The associated controls (regulators, filters, emergency/service control valves, pressure test points etc.) where required, shall be arranged in the correct sequence as indicated in Figure 5.

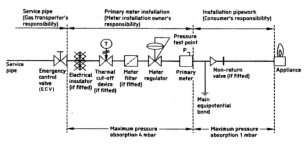

a) Low pressure (2nd family)

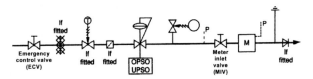

b) Medium pressure (2nd family)

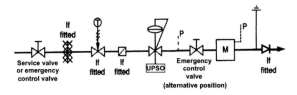

c) Low pressure (3rd family)

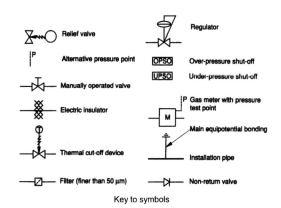

Key to symbols

Figure 5 Typical arrangement of meter and associated controls

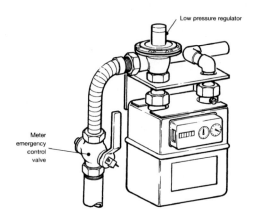

Figure 6 Low pressure meter installation

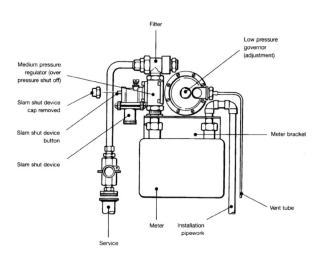

Figure 7 Medium pressure meter installation

Where pre-mix blown or compressed air or extraneous gas is to be used in conjunction with gas supplies, the gas meter and supply shall be protected by a non-return valve and the public gas transporter notified. The valve may be fitted at the meter outlet but, in certain cases, a non-return valve fitted at each appliance using air or an extraneous gas is an acceptable alternative.

It is not normally necessary to install a filter or regulator on a secondary meter.

Where electrical main equipotential bonding is necessary, a clamp should be used to make a connection to the outlet side of the primary meter. The connection must be as near as practicable and preferably not further than 600 mm of pipework from the meter outlet.

Conductors connected to the earth terminal should be of a size as laid down in BS 7671 Requirements for Electrical Installation (10 mm^2 cable with green and yellow insulation, construction reference 6491X conforming to BS 6004). See Figure 8 following.

Before any work is done that necessitates connection or disconnection of any meter or associated control, a temporary continuity bond shall be connected, by means of the clips or clamps, between the inlet of the emergency control valve and the installation pipe from the meter, ensuring that metallic contact is made. The temporary continuity bond shall remain in position until the work is completed.

WARNING NOTICES

A permanent notice shall be prominently mounted on or near a primary meter indicating to the gas user the action to be taken in the event of an escape of gas.

A similar notice is recommended on or near a secondary meter.

Where a service pipe supplies more than one primary meter fitted either in the same premises, or in different premises, the notices shall indicate how many meters are connected to the service pipe, and preferably their locations.

Where secondary meters are installed, the number and location of the secondary meters shall be mounted adjacent to the primary meter.

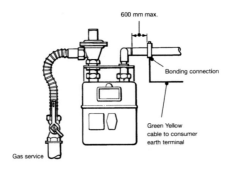

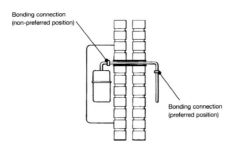

a) Built-in or surface mounted

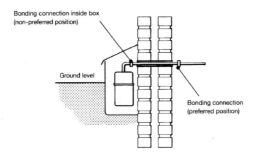

b) Semi-concealed

Figure 8 Bonding connection should be either inside property (preferred) or inside box (non-preferred)

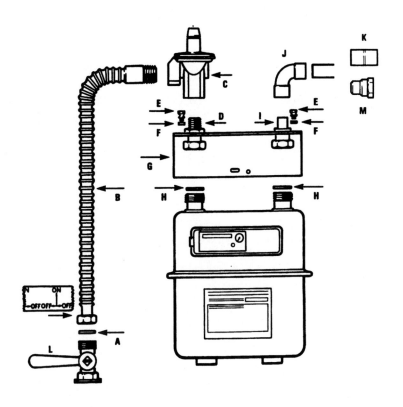

A ¾" meter washer
B Stainless steel connection
C Governor
D ¾" BSP taper grooved union
E Shear bolts
F Security washers
G Meter bracket
H Meter washers
I 22 mm grooved union
J 22 mm elbow
K 22 mm socket, copper internal
L ¾" gas tap
M ¾" connector, steel internal

Figure 9 Standard U6 primary meter components

TESTING AND PURGING

Before commencing any test the meter shall be visually inspected for signs of damage.

For 2nd family gases the installation shall be tested for tightness using either natural gas or air as the test medium, in accordance with IGE/UP/1B 2nd edition up to 35 mm diameter.

For 3rd family gases, prior to making any gas available, the installation shall be tested for soundness using air in accordance with BS 5482-1, 2 or 3 or BS EN ISO 10239.

TIGHTNESS TESTING FOR 2ND FAMILY GASES

Testing Procedures

Tightness testing gas pipework installations is achieved by means of a pressure test. A loss of pressure over a period of time indicates leakage from the installation.

The tightness testing procedure and results criteria will depend on whether the pipework is new or existing, up to 28 mm or 35 mm diameter and whether the appliances are new or existing. Any pipe installed for the first time and is therefore uncommissioned is deemed to be new pipework and must be tightness tested using natural gas or air as the test medium.

The procedures for tightness testing can be found in IGE/UP/1B 2nd edition or reference made to the ConstructionSkills publication *Gas Safety* (G1) September 2006 edition or later, as applicable. It should be noted that installations above 0.035 m^3 volume are outside the scope and should be tested using either IGE/UP/1A or IGE/UP/1.

Supporting Information for Testing where the Supply MOP Exceeds 75 mbar, But a Meter Inlet Valve (MIV) is Not Fitted

General

Current designs are to BS 6400 or IGE/TD/15 which require a MIV to be fitted. However, some earlier installations did not include a MIV and cannot be tightness tested to IGE/UP/1B 2nd edition. Guidance on the test procedure to be used can be found in IGE/UP/1B 2nd edition or ConstructionSkills publication *Gas Safety* (G1), September 2006 edition or later. It should be noted that where high pressure regulators are found to be faulty the installation should be upgraded to BS 6400-2 as opposed to simply replacing or repairing the regulator.

Selection of valves and their operational rating is dependent on the service supply pressure (e.g. some valves require a minimum service operating pressure of up to 500 mbar (0.5 bar).

The meter installer must ensure at first fix that the regulator to be installed is suitable to cover the range of pressures supplied (this should be checked at the design stage with the supplier before the installation is carried out).

Verification should also be made that the pipework and fittings between the ECV and MIV have been strength and gas tightness tested at the factory in accordance with IGE/UP/1 to withstand a design maximum incidental pressure (DMIP) of 2.7 bar and a maximum operating pressure (MOP) of 2 bar. **Note:** This may be in the form of a test certificate issued with the unit.

The service supply pressure should be checked by the meter installer with a suitable pressure gauge prior to commencing the installation (gauge should be capable of reading in excess of 2 bar).

SOUNDNESS TESTING FOR 3RD FAMILY GASES

Meter installations may be found in permanent dwellings, caravan park homes, mobile dwellings and permanently moored boats.

Testing of the installations may vary according to the situation and type of dwelling and reference should be made to the appropriate British Standard, e.g. BS 5482-1 and/or ConstructionSkills publication G18 *Liquefied Petroleum Gas* or G80 *Liquefied Petroleum Gas Safety* for details of the appropriate test/procedure to be undertaken.

RESULTS OF TIGHTNESS/SOUNDNESS TEST

Pass

Immediately after the installation has passed the tightness test it shall be purged with gas (see purging).

Fail

If the installation has failed any test then either:

a) the fault shall be traced and rectified, or

b) the emergency control valve or meter inlet valve outlet union shall be disconnected and sealed, or

c) the meter outlet shall be sealed and a notice shall be attached stating that the supply is not to be restored until the leak is repaired and the system satisfies the tightness test.

PURGING

The displacement of air by fuel gas or vice versa.

Gas Safety (Installation and Use) Regulations 1998, regulation 22(2), prescribes that every new and modified installation must be purged after passing the tightness test when connected to the gas supply.

Meters and installation pipework shall be purged to ensure that any explosive mixtures of gases present are removed. Natural gas burners need not be ignited during the purge procedure if the purge volume is less than 0.02 m^3, i.e. U6/G4/E6 meters with pipework not greater than 28 mm. For installations where U6/G4 meters include 35 mm diameter pipework and purge volumes above 0.02 m^3 then ignition should be applied during the purge.

During the purging procedure gas shall not be allowed to accumulate in confined spaces. Steps shall be taken to ensure good ventilation, particular care to be taken to ensure low level ventilation when purging heavier than air, 3rd family gases. Third family gas burners shall have a source of ignition at the burner during the purge process.

Operation of electric switches shall be avoided, and naked lights or smoking shall not be permitted in the vicinity.

Purging shall be conducted initially from the installation pipes furthest from the meter, after a successful tightness test has been completed.

The purge volume to be passed shall not be less than:

- 0.01 m^3 (0.35 ft^3) for G4 diaphragm/E6 ultrasonic (U6) meters with up to 28 mm diameter pipework

- refer to Gas Safety (G1) or IGE/UP/1B for any other meters and/or pipework up to 35 mm diameter.

Where appliances are already installed, all their burners shall be lit to ensure the purging has been completed.

Refer to the ConstructionSkills publication *Gas Safety* (G1) September 2006 or later or IGE/UP/1B for methods of calculation and further information.

METER AND REGULATOR OPERATION

After purging has been completed, the following procedure shall be implemented:

- any prepayment coin or token mechanism operation is checked

- the operation of the meter regulator is checked in accordance with any manufacturer's instructions. Regulators shall not be adjusted without the permission of the public gas transporter or gas supplier. The regulator working pressure may be checked using Table 2

- the regulator outlet pressure should be:
    - 28 ± 2 mbar for 3rd family butane gas
    - 37 ± 2 mbar for 3rd family propane gas

    In the case of 2nd family natural gas the pressure at the meter outlet may vary between 23 mbar and 19 mbar at flow rates between 0.5 m^3/hr and 6 m^3/hr. A flow rate of 3.5 m^3/hr with 30 mbar inlet pressure at the governor will result in 22 mbar at the governor outlet (when factory set) and should result in 21 mbar at the meter outlet. Pressures below 19 mbar should be investigated. However it should be noted that if a supply pressure of only 19 mbar is available at the outlet of the emergency control valve, due to excessive demand during the winter, the absorption across the meter could be up to 4 mbar if on full load (6 m^3/hr), giving a meter outlet reading of 15 mbar. In addition a 1 mbar drop across the system would result in the appliance inlet pressure being 14 mbar, which would be the minimum acceptable. Regulator lock up should not exceed 30 mbar.

- the regulator shall be adequately sealed so as to prevent its setting from being interfered with, without breaking the seal

- the meter index shall be incrementing correctly

- the meter index shall be read and recorded and the meter reading agency informed with the necessary details forwarded as appropriate.

Table 2 shows appliance rates for meter regulator settings. A minimum of one appliance should be operating when carrying out this procedure, which should result in a minimum flow rate of 0.5 m^3/hr as recommended. Note: remember this may result in up to 23 mbar gas pressure at the meter outlet.

Where an Ofgem Approved Meter Installer (OAMI) needs to adjust the regulator and no appliances or installation pipework exist, a meter check device, with an orifice to pass a gas rate of at least 0.5 m^3/hr, should be used ensuring that any gas is vented to the atmosphere in a safe manner.

Table 2 Appliance rates for meter regulator setting

Type of appliance	Rate
Central heating boiler	Full rate
Circulator	Full rate
Convector	Full rate
Flueless heater	Full rate
Gas cooker	3 burners on, full rate
Gas fire	Full rate
Hotplate boiling ring	Full rate
Instantaneous water heater	Full rate
Storage water heater	Full rate
Warm air unit	Full rate

If the meter regulator seal was broken then this must be replaced by the authorised installer when these checks are completed.

REMOVAL AND EXCHANGE OF METERS

A meter shall not be permanently removed without the authority of the owner.

When a meter is removed, the emergency control valve and the installation pipe at the meter shall be capped or plugged or otherwise permanently sealed.

Where a meter is used for billing purposes (whether primary or secondary/sub-deduct) the meter index reading shall be recorded, and along with any other details forwarded as appropriate.

Where a primary meter is removed and is not subsequently reinstalled or replaced by another meter, the person who removes the meter shall, before removal:

a) close any service valve that does not also control the gas supply to any other meter, and

b) clearly mark any live gas pipe in the premises in which the meter was installed to the effect that the pipe contains gas.

Where a meter exchange has been completed, the installation shall be tested for tightness, purged and the meter and regulator operation shall be confirmed as described previously.

BES PUBLICATIONS

Building Engineering Services continue to provide the gas, electric, water and refrigerant industries with a range of popular, respected and competitively priced publications.

These publications can be used either as the basis of training or for reference in the workplace. Some can also be used for assessment purposes. All are published in A4 format, with the most popular also available as A5, pocket-sized books.

DOMESTIC GAS

GAS SAFETY (G1)   Format: A4 in a ringbinder
The complete manual for reference or self-study. All of the essentials in 300 pages, with clear explanations and illustrations, covering ♦gas pipework ♦gas supply ♦combustion ♦appliance gas safety devices and gas controls ♦principles of gas flues ♦flueing standards ♦ventilation requirements ♦emergency procedures ♦unsafe situations ♦warning notices and labels. Also included is the HSE publication ♦*Safety in the installation and use of gas systems and appliances* (G31) which covers the HSE Gas Safety (Installation and Use) Regulations 1998 – Approved Code of Practice and Guidance, a ♦*Course Workbook* and a booklet of ♦*Practical Tasks* for you to complete.

GAS SAFETY (G2)   Format: A5 Wiro-bound
All the information and diagrams from *GAS SAFETY (G1)* in a handy size for reference on the job and for carrying in the service van.

DOMESTIC GAS APPLIANCES (G5)   Format: A4 in a ringbinder
Contains all seven of the domestic natural gas appliance manuals from ConstructionSkills in one package, plus the *Domestic Natural Gas Appliances Course Workbook (G14)*. The easy-to-use format makes it ideal for engineers working with a range of domestic appliances.
Each manual can also be purchased individually:
- Heating Boilers/Water Heaters (G7)
- Cookers (G8)
- Ducted Air Heaters (G9)
- Fires and Wall Heaters (G10)
- Tumble Dryers (G11)
- Meters (G12)
- Instantaneous Water Heaters (G13)

DOMESTIC GAS APPLIANCES (G6)   Format: A5 Wiro-bound
All the information and diagrams from the *DOMESTIC GAS APPLIANCES (G5)* in a handy size for reference on the job and for carrying in the service van.

FAULT-FINDING TECHNIQUES (G17)   Format: A4
Problems with locating that elusive fault? Follow the step-by-step techniques in this hands-on manual and speed up your fault finding on central heating systems.

SAFETY IN THE INSTALLATION AND USE OF GAS SYSTEMS AND APPLIANCES (G31)   Format: A4
An essential HSE publication for all those working with domestic gas. It gives advice on how to comply with *The Gas Safety (Installation and Use) Regulations 1998 – Approved Code of Practice and Guidance*, which has a special legal status. For example, if you are prosecuted for breach of health and safety law, and it is proved that you have not followed the relevant parts of the Code, a court will find you at fault (unless you can show that you have complied with the law in some other way).

COMMERCIAL AND INDUSTRIAL GAS

COMMERCIAL GAS SAFETY (G88)   Format: A4 in a ringbinder
An essential training and reference manual for those working in the commercial environment. It includes key sections from the popular GAS SAFETY (G1) and incorporates information from two other commercial publications (G23 and G24) which can be purchased separately) making this the definitive training and reference manual for commercial work. It covers ♦commercial gas safety ♦pipework and ancillary equipment ♦gas pipework ♦gas supply ♦combustion ♦appliance gas safety devices and gas controls ♦principles of gas flues ♦flueing standards ♦ventilation requirements ♦emergency procedures ♦unsafe situations ♦warning notices and labels. Also included is the HSE publication ♦*Safety in the installation and use of gas systems and appliances* (G31) and ♦*Course Workbooks* and *Practical Tasks* (G3, G4, G83 and G84).

COMMERCIAL GAS SAFETY (G23)   Format: A4
An essential supplement for engineers working in the commercial environment. If you already own a *GAS SAFETY (G1)* pack, all you need is this book with its commercial gas-specific sections ♦combustion and flue gas analysis ♦burners ♦controls and control systems ♦flues ♦ventilation ♦pressure and flow.

COMMERCIAL PIPEWORK AND ANCILLARY EQUIPMENT (G24)   Format: A4
An essential guide for engineers working on commercial pipework, with clear information on ♦pipework design ♦soundness testing and purging ♦commercial metering ♦boosters and compressors.

COMMERCIAL APPLIANCES (G25)   Format: A4
A comprehensive guide to the installation and commissioning of direct and indirect fired appliances, radiant heating and gas equipment.

COMMERCIAL CATERING (G26)   Format: A4
Essential information on installing, commissioning and servicing commercial catering appliances.

To obtain further information and order any of the publications listed, contact Publications
on: Tel: 01485 577800 / Fax: 01485 577758 / E-mail: publications@cskills.org / www.cskills.org/publications

LIQUEFIED PETROLEUM GAS (LPG)

LIQUEFIED PETROLEUM GAS SAFETY (G80) Format: A4 in a ringbinder/A4
The industry reference manual for those working only on LPG systems. It covers all you need to know about ♦combustion ♦appliance gas safety devices and gas controls ♦principles of gas flues ♦flueing standards ♦ventilation requirements ♦emergency procedures ♦unsafe situations ♦warning notices and labels.
This pack consists of: ♦*Gas Safety (G1) pack*, ♦*Liquefied Petroleum Gas Safety (G18) book*, ♦*Liquefied Petroleum Gas Safety Course Workbook (G81)*.

LIQUEFIED PETROLEUM GAS SAFETY (G18) Format: A4
The essential bolt-on to those working with natural gas and looking to extend into LPG. If you already own a *GAS SAFETY (G1)* pack, all you need is this book with its LPG-specific sections ♦installation ♦fire precautions and procedures ♦combustion ♦testing and commissioning installations ♦service pipework ♦bulk gas supply systems ♦the leisure industry.

ELECTRICAL

BS 7671: REQUIREMENTS FOR ELECTRICAL INSTALLATION (E1) Format: A4 Wiro-bound
The standard reference book for electrical work. The easy-to-follow text, supported by diagrams, explains the complex regulations in terms a practical electrician can understand. It now incorporates reference to the IEE on-site guide that enables you to make calculations and design circuits in a much quicker and simpler manner.

ELECTRICAL INSTALLATION PACK (E3) Format: A4 in a ringbinder
Over 430 pages of illustrated reference material divided into four sections:
- Basic Practical Skills – describes the tools required for electrical installation work and how to use them
- Wiring Installation Practice – deals with terminating cables, flexible cords and installing PVC cables, conduit trunking, MICC, SWA and FP200 wiring systems. (Complies with the 16th Edition *IEE Wiring Regulations*)
- Basic Electrical Circuits – covers standard circuit arrangements for lighting and power circuits, and relevant IEE Regulations
- Safety at Work – essential advice on safety at work, from securing ladders to dealing with electric shock. It also gives the key points of relevant Acts and Regulations.

ESSENTIAL ELECTRICS (E14) Format: A4
An indispensable reference book for plumbers, gas fitters and heating and ventilating engineers whose work requires basic electrical knowledge and an understanding of electrical regulations.

CENTRAL HEATING CONTROLS (E15) Format: A4
Deals with different types of central heating control systems for wiring and fault finding.

COMBINATION BOILERS (E19) Format: A4
An invaluable reference manual for engineers who want to understand the principles of combination boilers. This manual covers most of the content for the ConstructionSkills Essential Electrics and Combination Boiler Fault Finding course. Over 80 pages of illustrated reference information covering ♦types of boilers ♦designs ♦wiring diagrams ♦installation ♦commissioning and servicing ♦fault finding.

WATER

UNVENTED HOT WATER STORAGE SYSTEMS (W2) Format: A4
An informative guide for installing unvented hot water storage systems. It covers most of the content for the ConstructionSkills training and assessment scheme, including: ♦types of system ♦design ♦controls ♦installation ♦commissioning and decommissioning ♦servicing and fault diagnosis ♦relevant Building Regulations ♦good practice.

REFRIGERANTS

SAFE HANDLING OF REFRIGERANTS (R2) Format: A4
Essential information, primarily designed for operatives undertaking ConstructionSkills Safe Handling of Refrigerants training and assessments, it covers ♦environmental impact ♦fluorocarbon control and alternatives ♦regulations ♦recovery and handling ♦refrigeration theory ♦good practice ♦automotive installations.

SAFE HANDLING OF ANHYDROUS AMMONIA (R4) Format: A4
Essential information for handling anhydrous ammonia. Primarily designed for operatives undertaking ConstructionSkills Safe Handling of Anhydrous Ammonia training and assessments, it covers ♦safety and environmental issues ♦regulations ♦good practice.

PIPEWORK AND BRAZING (R6) Format: A4
Primarily for operatives undertaking ConstructionSkills Pipework and Brazing training and assessments for refrigeration systems, it covers ♦health and safety ♦materials and equipment ♦lighting procedures.

To obtain further information and order any of the publications listed, contact Publications on: Tel: 01485 577800 / Fax: 01485 577758 / E-mail: publications@cskills.org / www.cskills.org/publications